ROYAL COMMISSION

ON

ENVIRONMENTAL

POLLUTION

CHAIRMAN:
THE RT HON THE LORD LEWIS OF NEWNHAM

FIFTEENTH REPORT

EMISSIONS FROM HEAVY DUTY DIESEL VEHICLES

Presented to Parliament by Command of Her Majesty
September 1991

HMSO: LONDON
£12.00 net

Cm 1631

Recycled Paper

Previous Reports

ROYAL COMMISSION ON ENVIRONMENTAL POLLUTION

FIFTEENTH REPORT

To the Queen's Most Excellent Majesty

MAY IT PLEASE YOUR MAJESTY

We, the undersigned Commissioners, having been appointed "to advise on matters, both national and international, concerning the pollution of the environment; on the adequacy of research in this field; and the future possibilities of danger to the environment";

And to enquire into any such matters referred to us by one of Your Majesty's Secretaries of State or by one of Your Majesty's Ministers, or any other such matters on which we ourselves shall deem it expedient to advise:

HUMBLY SUBMIT TO YOUR MAJESTY THE FOLLOWING REPORT.

"This most excellent canopy, the air . . .
why, it appears no other thing to me
but a foul and pestilent congregation of vapours."

Shakespeare, Hamlet, Act II

"This City now doth, like a garment, wear
The beauty of the morning; silent, bare,
Ships, towers, domes, theatres, and temples lie
Open unto the fields, and to the sky;
All bright and glittering in the smokeless air."

Wordsworth, Composed upon Westminster Bridge, 1807

PREFACE

This Report is the product of a new way of working on the part of the Royal Commission. In July 1990 the Commission announced that, alongside its major studies, it would undertake a series of short studies on tightly focused topics. These would enable it to cover a wider range of environmental issues at any one time and to respond more quickly as issues arose. The Commission decided that the first such study should be on the environmental pollution caused by emissions from heavy duty diesel vehicles.

Having defined more precisely the scope of the study, the Commission established a group of members to take it forward. The group was chaired by Mr William Scott and included Professor Henry Charnock, Professor Dame Barbara Clayton, Lord Lewis and Professor Aubrey Silberston. The group presented an interim report of its conclusions, which was reviewed by the full Commission, and then prepared the draft of the final report. The draft was considered three times by the full Commission and the resulting Report is issued as a report of the Royal Commission.

CONTENTS

INFORMATION BOXES

TABLES

FIGURES

PLATES

Plates 1-4 are located after page 4 and Plates 5-9 after page 36.

CHAPTER 1

SCOPE AND EMPHASIS OF THE REPORT

Importance of the Topic

1.1   In the Tenth Report([1]) the Commission commented on air quality, including the contribution made by motor vehicles to air pollution. It welcomed the attention being paid to the environmental, technical and economic aspects of vehicle emissions by the Commission of the European Communities and by other bodies. It did not attempt, in that Report, to reach a definitive view on the extent of further reductions in emissions which was desirable, which pollutants required priority attention and by what technical means any reductions could be achieved. It did, however, express its intention to keep these issues under review.

1.2   That was in 1984. Since then the number of vehicles and the total distance travelled by them have continued to increase in the UK and world-wide. In the UK, between 1984 and 1989, there was estimated to be an increase of about 33% in the distance travelled by all vehicles. There was also an estimated increase of about 33% for goods vehicles but a quadrupling for the largest of them([2]). Road vehicles have become the largest source of nitrogen oxides in the UK, with diesel vehicles contributing a large and growing proportion. Diesel vehicles are also the major source of smoke in urban areas([3]).

1.3   The attention paid to the topic of air pollution from vehicle emissions by the European Commission and others has led to successively tighter emission limit values being required of new vehicles. Passenger cars have been the main focus of attention. Until recently heavy duty vehicles have been controlled only in respect of their emissions of smoke, and that not very tightly. The first European Community Directive to control gaseous emissions from heavy duty diesel vehicles came into force in October 1990([4]). The European Commission has now proposed an amendment to it, tightening the limits and introducing a limit for particulate emissions; a text has been agreed by the Council of Ministers([5]).

Scope of the Study

1.4   This Report considers emissions from heavy duty diesel vehicles. The term 'heavy duty vehicle' has no one precise meaning. For this study we took it to refer to a goods vehicle of more than 3.5 tonnes gross vehicle weight or a passenger vehicle with 6 or more seats. This is a commonly accepted definition and is the one used for the EC Directive referred to above. More than 90% of the heavy duty vehicles in this country, and virtually all of the heaviest vehicles, are powered by diesel engines.

1.5   The study has focused on:

> future emission standards, with particular reference to the approach which should be taken for standards to be implemented in the European Community at the end of the decade;

> means of abatement of emissions, including the scope for reducing emissions from vehicles in service;

> the development and enforcement of emission standards to be met by vehicles in service; and

the significance for emissions of diesel fuel characteristics and of additives and lubricants.

In addressing these issues, account has been taken of the possible use of economic instruments to achieve policy objectives.

1.6 The study has excluded important aspects such as emissions of noise and of carbon dioxide. Both deserve attention but could not be covered within the timescale we sought for completion of this study. It is worth commenting briefly on the very different stages of development of control of these and their relationship with the control of other emissions.

(a) Vehicle noise is already subject to an EC Directive[6] and there are substantial programmes of research and development in the UK[7] and in many other countries to design quieter vehicles. We were pleased to learn that developments in the control of the emissions we have focused upon, notably nitrogen oxides, are expected to contribute also to reductions in engine noise[8].

(b) Emissions of carbon dioxide, by contrast, are at present subject to no control. National Governments and international bodies, including the European Community, are considering what form of control would be feasible. The level of carbon dioxide emissions from vehicles is linked closely to their fuel consumption. Some of the measures required to control other emissions will have the effect of slowing the trend towards increased efficiency in the use of fuel in vehicles and of energy in oil refining, leading to higher emissions of carbon dioxide than would otherwise have been the case. If this effect were to become significant in respect of any such measure it would need to be taken into account in judging its merits.

1.7 Whilst its focus has been as described above, the study has raised a number of other, related, issues. Emissions from off-road uses of diesel engines and the prospects for use of alternative fuels are considered briefly in the Report. Emissions from light duty diesel vehicles such as taxis are not addressed directly but some of our recommendations could be applied also to them.

1.8 Other issues, such as the merits and feasibility of switching between modes of transport and of imposing some kind of restraint on the growth in numbers and use of vehicles, are not addressed. They are important, and should be taken into account in determining national transport policies, but they are too wide-reaching to include within this study. We note an enhanced awareness of environmental issues indicated in recent statements of the Government's transport policies[9,10].

1.9 The objective of our study has thus been to consider pollution from, and means of securing reductions in, emissions from heavy duty diesel vehicles. Most of our recommendations, if implemented, would lead to more effective control of emissions from all such vehicles. As explained in paragraph 1.11, however, we have made additional recommendations in respect of certain classes of vehicle.

Contents of the Report

1.10 The Chapter which follows considers the emissions with which we are chiefly concerned, how they are formed within the engine, how they may be controlled and what impact they have on the environment. It is supplemented by a technical paper prepared for us by Ricardo Consulting Engineers Ltd and reproduced as Appendix 4. Chapter 3 considers issues related to the standards of emissions control to be achieved by new vehicles in the European Community, including the basis for emission limit values and the choice of engine test cycle. Chapter 4 turns to vehicles in service, considering what emissions standards should be expected of them and the scope for reducing emissions by retrofitting appropriate equipment or by other means. Chapter 5 considers the role of diesel fuel in emissions control, the role of additives and lubricants and, briefly, the scope over the next decade for using alternative fuels. The thematic summary at Chapter 6 sets out some of the key topics of the Report, bringing together aspects which are made in one or more of the earlier chapters. Two such topics — protection of the urban environment and the application of economic instruments — are addressed below.

The Urban Environment

1.11   Our objective is to ensure protection of the environment in all parts of the country and most of our recommendations are intended to result in improvements wherever diesel emissions may be a problem. Some parts of the country, such as the more heavily used motorways and the larger urban areas, are subject to very high concentrations of vehicles and their emissions. In addition, the number of people and buildings exposed to the pollutants is greater in urban areas than elsewhere. We have therefore considered whether additional steps to control emissions in urban areas might be warranted. We have reached the following conclusions.

(a)   The main contribution to emissions from heavy duty vehicles in urban areas is from vehicles, or classes of vehicle, which operate in all parts of the country. Attention should therefore be directed towards the control of emissions from all vehicles, with an emphasis on the control of emissions generated by them under urban driving conditions - that is, at low and medium engine speeds and at relatively high engine loading. This is considered further in Chapter 3 in the context of the engine test cycle.

(b)   Special attention should be directed towards the control of emissions from vehicles which operate mainly in urban areas. We have particularly considered buses, including those, such as school buses, which do not normally carry fare-paying passengers. In addition to driving in urban traffic, with its low average road speed and frequent changes in engine speed, buses are required to make frequent scheduled stops and starts. This poses a stiff challenge for the control of emissions, especially of particulates. Also, many buses are kept in operation for substantially more years than are most heavy goods vehicles, so that improvements in the emissions control standards achieved by new vehicles will spread only slowly through the whole fleet. The implications of this for emissions control are considered in Chapters 4 and 5.

(c)   Many of the considerations which apply to buses apply also, to varying extents, to other classes of vehicle which operate mainly in urban areas. Refuse collection vehicles are a notable example. We recommend that the Government should carry out the necessary work to identify the classes of vehicle which make the largest contributions to emissions in urban areas and should consider how our recommendations for buses might be applied to them.

(d)   In addition to technical means for controlling emissions, measures to reduce the extent of vehicle use in urban areas could contribute to a decrease in emissions there and an improvement in air quality. Such measures should be considered. They raise broader issues, however, and do not fall within the scope of this Report.

Economic Instruments

1.12 We consider that the control of emissions from vehicles, including heavy duty diesel vehicles, is an appropriate area for the application of economic instruments: that is, the creation by Government of financial incentives of some kind. There are two reasons for this.

(a) *Pollution results from the cumulative impact of large numbers of vehicles.* An economic instrument which influenced a substantial proportion of operators to use less polluting vehicles, even though a minority persisted with relatively more polluting ones, could improve the overall position significantly. It could do so more cost-effectively than reliance upon regulation based on universally applicable standards. Universal standards impose higher costs on some operators than on others. The standard is generally chosen so that few if any operators face costs which are considered (by the authorities) to be unreasonably high. Many operators could afford to meet a higher standard without incurring high costs. An economic instrument could encourage them to do so, thus reducing the total of emissions.

(b) *There could be more rapid development and use of new technology.* Emissions control technology is developing continuously, partly in response to regulation. New standards are set, on the basis of what appears to be technically feasible, to be met by a specified date. There is always an element of uncertainty as to whether the chosen date is the earliest practicable. If it is chosen so that most manufacturers are able to meet it, some will probably be able to achieve the set standard earlier. An economic instrument could encourage such early achievement, securing a quicker improvement in total emissions. It could, in principle, also provide a continuing incentive for the replacement of existing vehicles or engines with new, less polluting ones.

1.13 There is scope for the application of economic instruments to the control of emissions from new vehicles and from new or improved engines for vehicles in service. They may also be applied to the sale of fuel. These three areas are considered in Chapters 3, 4 and 5 respectively.

Plate 1(a) Buses and heavy goods vehicles in central London.
Photograph by courtesy of Environmental Picture Library/Vanessa Miles

Plate 1(b) A lorry emitting visible smoke.
Photograph by courtesy of Transport and Road Research Laboratory

(a)

(b)

Plate 2(a) A turbocharger.
Plate 2(b) A turbocharger on an engine, with aftercooling.

Exhaust gas (1) from the cylinders drives a turbine (2). Linked to it is a compressor (3) on the air intake. The compressed air passes through an aftercooler (4) before entering (5) the cylinders.

Photograph by courtesy of Perkins Technology Ltd

(a)

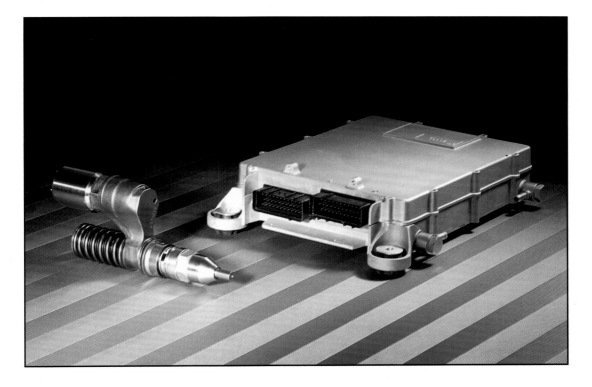

(b)

Plate 3   Advances in fuel injection equipment:-

(a)   An in-line pump manufactured in 1934. A set of individual pistons, under mechanical control, supply measured amounts of fuel at the correct timing. The fuel is pumped to fuel injectors on each cylinder (not shon).

(b)   An electronic unit injector system. A separate fuel pump combined with each injector, one of which is shown alongside the electronic control unit which manages the whole fuel injection system. Unit injectors are poositioned in the head of each cylinder. Much higher pressures and more precise control of injection are obtained in this way.

Photograph by courtesy of Lucas Powertrain Systems

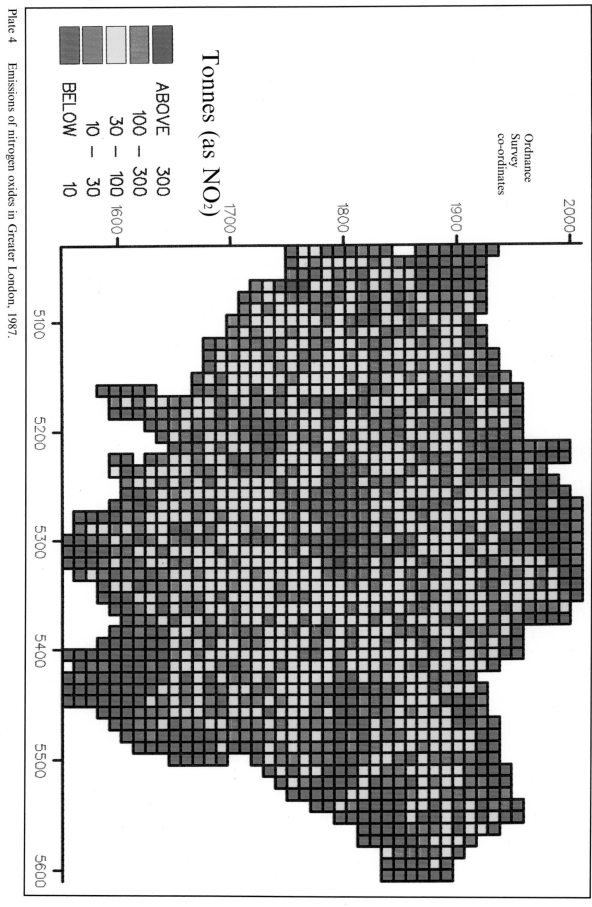

Plate 4   Emissions of nitrogen oxides in Greater London, 1987.

Source: NAEI, Warren Spring Laboratory[24].

CHAPTER 2

EMISSIONS

Types of Emission

The Diesel Engine

2.1 The diesel engine was invented in 1892 by Rudolf Diesel. Its operation depends upon the ignition of fuel vapour by the heat of compression of the air in the cylinder. The end products of combustion discharged in the exhaust are carbon dioxide, carbon monoxide, oxides of nitrogen, unburnt hydrocarbons (including polyaromatic hydrocarbons), particulate matter and water vapour. The particulate matter is made up of insoluble material such as carbon particles, ash and adsorbed hydrocarbons, together with water soluble material such as sulphates and nitrates. The operation of the diesel engine is described in the box overleaf and more fully in Appendix 4 section 2.

2.2 A heavy duty, direct injection diesel engine has lower fuel consumption than a petrol engine of equivalent power, even after the effects of fuel density are taken into account, so that its emissions of carbon dioxide are lower. Emissions of hydrocarbons and carbon monoxide are also lower from a diesel engine but emissions of nitrogen oxides are somewhat higher and of particulate matter very much higher. A petrol engine fitted with a three-way catalyst, however, has lower emissions of nitrogen oxides and hydrocarbons than the diesel engine whilst emissions of carbon monoxide are about equal.

2.3 Concern about the emissions of the diesel engine centres on the oxides of nitrogen, unburnt hydrocarbons, visible smoke and particulate matter. Together with carbon monoxide these emissions are the subject of legislative control, as described in the next Chapter. This Chapter describes how the emissions are formed in the engine, how they may be controlled, the total amounts emitted and the impact that they have on human health and on the natural and the built environments.

Diesel Emissions

2.4 Key points about diesel emissions are described below. Appendix 4 section 3 gives further details.

> *Nitrogen Oxides (NOx).* NOx refers generically to a number of compounds of nitrogen and oxygen. The principal compounds found in diesel emissions are nitric oxide (NO) and nitrogen dioxide (NO_2). Nitrogen oxides arise in the diesel engine from the effect of the high combustion temperature on nitrogen and oxygen in the air drawn into the cylinder.

> *Hydrocarbons.* Diesel fuel consists largely of a mixture of many different hydrocarbons, compounds whose molecules contain hydrogen and carbon. Some of these molecules contain chains of carbon atoms which may be straight or branched. Some, known as aromatics, contain carbon atoms in structures known as benzene rings. Much of the fuel is completely burnt to produce carbon dioxide and water. Some fuel, however, typically 0.3% in a modern engine, undergoes only partial combustion and gives rise to hydrocarbon molecules with shortened carbon chains and small amounts of oxidised compounds such as alde-

THE OPERATION OF THE DIESEL ENGINE

The diesel engine resembles the petrol engine in several major features. A piston moves up and down in a cylinder which forms the combustion chamber. Air is drawn in, and exhaust expelled, through inlet and outlet valves. The diesel differs from the petrol engine, however, by working on a compression ignition cycle, mixing air and fuel in the cylinder rather than in a carburettor and igniting the fuel vapour by the high temperature resulting from the compression of air rather than by a spark plug. The diesel is a lean burn engine, that is, an engine which works with an excess of air over the minimum that is required for the complete combustion of the fuel.

The great majority of diesel engines mounted in vehicles operate on a four stroke cycle; this is illustrated in Figure 2.1. As the piston travels downwards at the start of the *intake stroke* (1), air is drawn into the cylinder through the open inlet valve. The piston reaches the bottom of its stroke and starts moving back upwards again, the inlet valve closes and air is trapped in the cylinder. As the *compression stroke* (2) continues, the air is compressed to more than 45 atmospheres and heats up to more than 500°C (as hot as a ring on an electric cooker at maximum setting). Towards the end of the compression stroke fuel is pumped into the cylinder in a fine spray through an injector nozzle. The fuel droplets rapidly mix with the air and ignite. At the very high flame temperature some of the nitrogen in the air also reacts with oxygen to form nitrogen oxides (NOx). The temperature of the gases rises sharply, increasing the pressure and causing the gases to expand. This drives the piston downwards on the *expansion stroke* (3), delivering power to the crankshaft. Injection continues for a short time at the start of the expansion stroke. Fuel droplets which move into regions that are still rich in oxygen are able to burn completely. Other fuel droplets, if surrounded by insufficient oxygen, burn only partially. Where oxygen is absent fuel droplets fail to burn and are pyrolysed by the high temperature. This leads to the appearance of unburnt hydrocarbons and particulate matter in the exhaust. As the piston reaches the bottom of the cylinder the exhaust valve opens. The burnt gases are forced out in the final *exhaust stroke* (4) as the piston moves upwards under the momentum of the crankshaft and flywheel. On completion of the exhaust stroke the cycle is set to start again.

There are two major types of diesel engine. In the Direct Injection (DI) engine the fuel is injected directly into a bowl within the crown of the piston. The Indirect Injection (IDI) engine has a separate antechamber where fuel mixes with air before entering the cylinder. IDI engines are able to work over a wider speed range than DI engines and are more suitable for small scale applications. They are also less noisy. There is, however, some loss of efficiency in the IDI engine. For these reasons the DI engine has, until now, been used on larger vehicles and the IDI engine for smaller, high speed engines such as those fitted to cars and light vans. It is expected, however, that development of the DI engine will lead to its wider use in smaller vehicles.

hydes. Hydrocarbons emitted in the exhaust are present as a gas and as a component of particulate matter (see below). Aromatic molecules may react together and the benzene rings join to form polyaromatic hydrocarbons (PAHs), some of which are carcinogens. A small proportion of the fuel remains unburnt and passes into the exhaust. Lubricating oil which enters the combustion chamber also contributes to the hydrocarbon component associated with particulate matter.

Smoke. Incomplete combustion of the fuel may result in the formation of carbon particles. On formation most of the particles are very small (less than 1 micrometre across) and invisible individually but, whilst

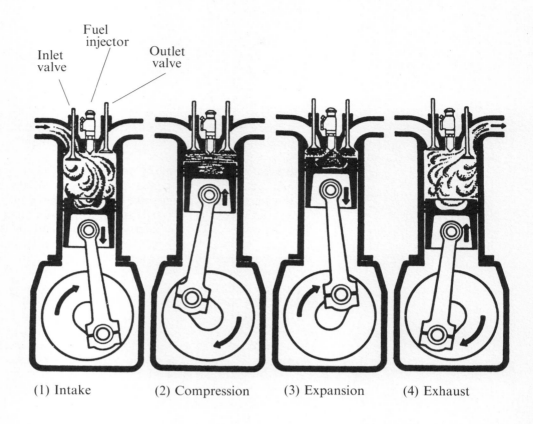

Inlet valve Fuel injector Outlet valve

(1) Intake   (2) Compression   (3) Expansion   (4) Exhaust

Figure 2.1 Operation of a diesel engine.

moving from the engine to the atmosphere, some of the particles aggregate. At high concentrations, particles are seen in the air as black smoke. White and blue smoke, formed by droplets of unburnt fuel and lubricating oil, may be emitted when the engine and the ambient air are cold.

Particulate matter. Total particulate matter is measured by collection on a filter under standard conditions. The large total surface area presented by particulates enhances the adsorption of hydrocarbons, sulphates and water. Although the composition of particulate matter varies widely, a chemical analysis of a typical sample found 41% elemental carbon, 25% derived from lubricating oil, 7% from fuel, 14% sulphate and water, and 13% other material which is mostly metal from engine wear and inorganic residues from additives[11].

Sulphur Dioxide and Sulphates. Diesel fuel contains sulphur in small quantities which is oxidised to sulphur dioxide. Most sulphur is emitted in that form but a limited amount, typically 2%, is further oxidised to sulphur trioxide which reacts with water to form sulphuric acid. Some of the sulphuric acid combines with trace metals and organic material to form sulphates. These, together with droplets of sulphuric acid aerosol, are all categorised as particulate matter in the engine test.

Carbon Dioxide and Carbon Monoxide. Excess air is usually available in the combustion chamber and the great proportion of the fuel is fully burnt to carbon dioxide and water vapour. Incomplete combustion of some of the fuel, under conditions of full load when the air to fuel ratio is low, results in the emission of small amounts of carbon monoxide.

Control of Emissions

Engine Technology

2.5   Improvements to the diesel engine have already done much to reduce emissions. Heavy duty engines manufactured about ten years ago commonly produced NOx emissions of the order of 15–25 grammes per kilowatt hour (g/kW.h: this unit is explained in paragraph 3.8). Engines which comply with the present EC Directive (paragraph 3.5) emit no more than the lowest point of this range. Further improvements to meet the emission limit values for NOx agreed by the European Community for 1996 (paragraph 3.7 and Table 3.1) will reduce this to less than 7 g/kW.h. The extent of the reduction in the emission of particulate matter is planned to be even greater: emissions from current engines are often around 0.6–1.0 g/kW.h but will be 0.15 g/kW.h or less for engines manufactured after 1996.

2.6   The load and rotational speed of the engine, and the design of the combustion chamber, are key factors influencing emissions. For heavy duty engines:-

NOx is produced at very low levels at low loads, increasing towards a plateau figure at higher loads which are generally associated in typical operating conditions with higher engine speeds.

The formation of particulate matter is the result of several complex processes and has no set relationship with operating conditions. There is a tendency, however, for particulate formation to be greatest either at high loads and lower speeds, when the carbon fraction is produced, or at low loads and higher speeds when the hydrocarbon fraction is produced.

2.7   Strategies for reducing emissions from heavy duty vehicles have paid most attention to better design of the engine. Low emission engines rely on the precise control of the complex process of combustion of diesel fuel. The mixing of air and fuel can lead to local regions in the combustion chamber which are rich in fuel but sparse in air so that incomplete combustion results (see box on page 6). Much effort is therefore directed towards engine designs which deliver an adequate charge of air at all stages of operation and which produce thorough mixing of the fuel with the air. The greatest advances have been made by redesigning the combustion chamber, improving the injection of fuel and increasing the charge of air by means of turbocharging and aftercooling. These are described in more detail in Appendix 4 section 4. Many of these developments entail extra cost as they require either additional equipment, such as turbochargers and aftercoolers (Plate 2), or the replacement of equipment such as fuel injectors (Plate 3) by systems which are more complex and are manufactured to higher engineering standards.

Exhaust Aftertreatment

2.8   Exhaust aftertreatment holds out the possibility of compensating for the limitations of engine improvements, especially in controlling particulates.

PARTICULATE TRAPS

The emission of particulate matter may be controlled by filtering the exhaust gases through a particulate trap. In one design that is currently on trial the gases pass through fine channels in a porous ceramic monolith (Figure 2.2). The channels of the monolith are sealed at one end and the gases are forced through pores in the channel walls, trapping the particulate matter. Other designs incorporate a ceramic foam or a ceramic yarn as the filter element. As the filter would rapidly become blocked there must also be a means of clearing or regenerating it. This is done by burning off the particulate matter which has been deposited.

The gases of diesel exhaust are too cool to burn off the particulate matter directly, except under conditions of full load. Higher temperatures may be obtained by restricting the flow of the exhaust gases or by installing a supplementary heater powered by diesel fuel or by an external electricity supply. A fuel penalty of about 1% of the vehicle's consumption is incurred by this[12]. The trap exerts a back pressure on the exhaust stream which rises as particulate matter builds up in it. Regeneration is initiated when a certain pressure is reached. The monitoring and control system required for this makes such traps complex and expensive devices.

The ignition temperature of the particulate matter may be reduced by coating the filter with a metal catalyst, such as copper, or by adding a catalyst to the fuel. This reduces, and may eliminate altogether, the need for regeneration equipment. One fuel additive trap is currently undergoing trials in the UK[13]; it requires no regeneration equipment and takes up little space. Appendix 4 section 5 gives further details of particulate traps.

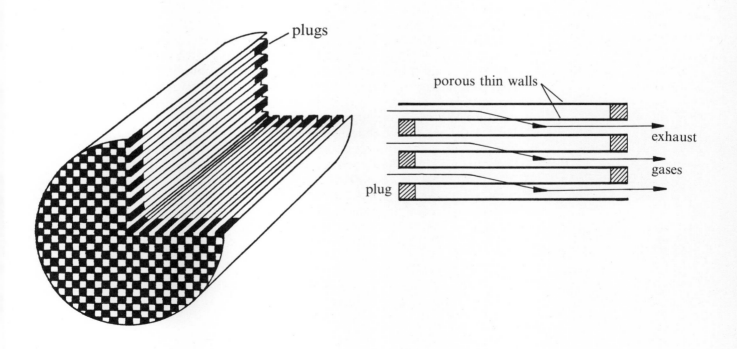

Figure 2.2 Schematic diagram of a particulate trap.

2.9 Particulate traps, whose operation is described in the box on page 9, are on operational trial in many parts of the world. Most models are bulky (see Plate 7) and cannot easily be installed on certain vehicles, such as articulated tractive units, because of space limitations[14]. The cost varies according to the design: we are aware of one example of a trap costing £3,000[15] and another costing £10,000[16], both suitable for fitting to large vehicles. Costs are expected to fall with longer production runs but it is not clear by how much. Questions over the reliability of the control systems and lifetime of the trap material have still to be resolved, although extensive trials in Germany provide some encouraging indications. For traps to be included in the German trials the average efficiency in separating particles must be shown to be better than 70%, or 50% for engines which do not exceed emissions of 0.4 g/kW.h. Some exceed 90% efficiency[16].

2.10 Research is also being directed towards the development of a flow-through catalyst which would substantially decrease the emissions of gaseous hydrocarbons and the hydrocarbon component of particulates by oxidising them to carbon dioxide and water[17]. A flow-through catalyst is simpler and less bulky than most particulate traps, as it does not require equipment for regeneration, and installation costs are likely to be less. It has previously been thought that a catalyst which would chemically reduce the oxides of nitrogen, such as operates in three-way catalyst convertors fitted to petrol engines, could not operate in the diesel exhaust stream because of the rich supply of oxygen. We understand, however, that such a device is now being developed[18], though it will be several years before its use becomes practical.

Further Developments in Emissions Control

2.11 Major advances in engine design have been made in recent years. Further development is expected but there appear to be few radically new technical options. Most promise is held out by the wider application of electronic monitoring and management, especially of the precise timing and rate of fuel injection. There may eventually be a limit to the gains which can be achieved by the technical improvement of engines themselves because of the intrinsic physics of their design. In the opinion of one manufacturer there is a natural barrier to decreasing NOx emissions below 4–5 g/kW.h and particulate matter below 0.1 g/kW.h[19]. Exhaust aftertreatment devices are likely to be necessary to achieve reductions below these levels. They may be expected to become increasingly effective. The potential impact of changes in fuel characteristics, which have previously been marginal, are also becoming proportionately more significant; we expect this to be reflected by a shift in the emphasis of development. Efforts to develop engines with even lower emissions will look beyond the diesel to novel engines such as diesel-electric hybrids, to engines which do not operate on the compression ignition principle and to the use of alternative fuels.

Energy Efficiency

2.12 One way to lower emissions from a vehicle is to increase the engine's efficiency in the use of fuel. Some means of increasing the power output per unit of fuel consumed are described in section 4.5.2 of Appendix 4. Another way is to reduce the power required to propel the vehicle, thus allowing a less powerful and therefore less polluting engine to be used. This can be achieved by improving the aerodynamics and the transmission system of the vehicle. Fuel economy trials have shown possible improvements of up to 23% by fitting bodies with low air resistance. There has been a trend in this direction since the 1960s[20] and improvements are expected to continue. Speed limiters for heavy goods vehicles, similar to those already fitted to coaches, will be an additional benefit. In the short term they will reduce the power expended by existing engines and in the longer term may encourage the use of less powerful engines. We welcome the Government's proposal to introduce speed limiters for heavy goods vehicles([3]).

Trade-Offs between Emissions

2.13 Although it may be the engineer's aim to reduce all classes of emission, in practice technical solutions often involve a trade-off in which a decrease in one emission leads to an increase in another. The most notable of these is the trade-off between the emission of NOx and the emission of particulate matter which is discussed below. Other trade-offs must, however, be made at the same time. Control of NOx is achieved at the expense of less efficient combustion; this leads to a rise in fuel consumption and the production of more carbon dioxide. A significant proportion of hydrocarbon emissions have their origin in lubricating oil; reducing the consumption of oil can cut emissions but may increase wear and reduce the lifetime of the engine.

2.14 *The Trade-Off between NOx and Particulate Matter.* Engine improvements that lead to decreased emissions of NOx mainly involve reducing the combustion temperature. In contrast, effective control of emissions of particulate matter requires faster, higher temperature combustion. At a given load, combustion temperature is largely determined by the quantity of fuel in the cylinder at the start of ignition. The rate of supply of fuel relative to combustion and the piston cycle is controlled by the timing of injection (Appendix 4 section 4.2.2). If the injection of fuel is retarded until later in the compression stroke (see box on page 6), less fuel is mixed with air before combustion and the temperature and pressure of initial combustion are reduced. This results in the production of less NOx from the air. A higher proportion of the fuel is injected later in the cycle, however, resulting in the formation of more particulate matter. The trade-off between the production of NOx and of particulate matter is a fundamental feature of engine operation and is influenced by many aspects of engine design as well as the setting of the injection timing. The trade-off is illustrated in Figures 2.3 and 2.4. Figure 2.3 shows, for each of three engine types in current production in Europe, variations in the production of NOx and of particulate matter in response to adjustments to injection timing and aspects of engine design. The trade-off curves for three prototype engines are shown in Figure 2.4.

2.15 It can be seen from these Figures that, for any given engine type, an attempt to achieve very tight control of either NOx or of particulate matter can be successful only by allowing the other class of emission to increase disproportionately. Alternatively, a moderate degree of control of both NOx and particulate matter together may be achieved. Lower levels of both NOx and particulate matter may be obtained by the application of a different control technology. The Figures indicate the increasing degree of control that is achieved by the use of turbocharging, turbocharging with aftercooling and turbocharging with aftercooling plus exhaust gas recirculation.

2.16 In practice, however, engine design is not the only factor that influences the trade-off between NOx and particulate matter. An alternative line of development is to look to engine improvements to control NOx, while using exhaust aftertreatment to remove hydrocarbons and particulate matter. The more stringent emission limit values agreed for the European Community in 1996 (Table 3.1) will require a package of further improvements in engine design and the use of low sulphur fuel. It is generally considered within the engine industry that, with such fuel, medium duty and heavy duty engines fitted with turbochargers or turbochargers and aftercooling will be

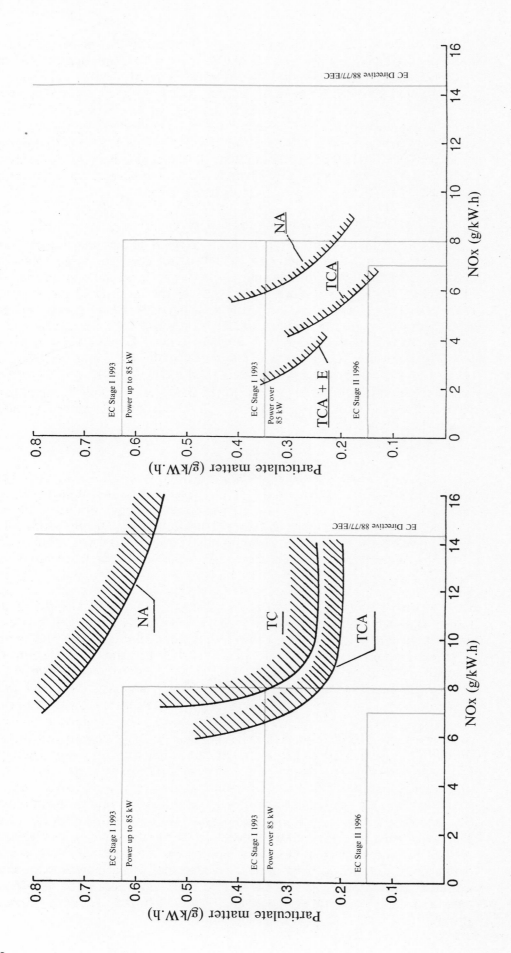

Figure 2.3 Trade-off between particulate matter and NOx: current engines

Figure 2.4 Trade-off between particulate matter and NOx: advanced research engines

Source: Ricardo Consulting Engineers Ltd.

Each curve under the hatched lines indicates the boundary of the scatter of measurements of emissions of NOx and particulate matter made in different engine types. Figure 2.3 shows results from several current European production engines of 3 types; Figure 2.4 shows results from 3 prototype engines using advanced emission control technologies. Measurements were made over the European test cycle as described in paragraph 3.22 and the box. The engines types represented are: naturally aspirated engines (NA), turbocharged engines (TC), turbocharged with aftercooling (TCA), and turbocharged with aftercooling plus exhaust gas recirculation (TCA+E). These are described in sections 4.1.1, 4.1.2, 4.1.3 and 5.3 respectively of Appendix 4. The limit value for NOx specified in EC Directive 88/77/EEC is shown as a vertical line; there is no limit for particulates in that Directive. The limit values for emissions of NOx and particulates agreed for the European Community in 1993 and 1996 (Table 3.1) are shown for comparison.

capable of meeting those values. It has yet to be resolved, however, whether naturally aspirated engines will be able to meet them at an acceptable unit cost without the addition of exhaust aftertreatment devices. The engine could be designed so as to achieve greater control of NOx if the requirement for it to control particulates were to be relaxed as the result of fitting aftertreatment for particulate matter. Exhaust gas recirculation, the trade-off curve for which is shown in Figure 2.4, provides such a technology although its drawbacks in terms of increasing wear on the engine have yet to be over-come. Similarly, the engine could be designed to achieve greater control of particulates if aftertreatment of NOx were available.

The Amount and Fate of Emissions in the Environment

2.17   In 1989 there were 567,000 diesel powered goods vehicles of more than 1.5 tonnes unladen weight and 103,000 diesel powered 'public transport vehicles' (including taxis) of all sizes, licensed for use on the road in the UK (Table 2.1). It has been estimated that 82% of the total diesel fuel used by road vehicles in the UK in 1988 was consumed by heavy goods vehicles (HGVs) (more than 3.5 tonnes gross weight) and 4% by passenger service vehicles (PSVs); the remainder was used by vans, cars and taxis[21]. There has been some growth in the last few years in the total number of diesel vehicles and more marked growth in the total distance travelled by them and in the consumption of diesel fuel. Growth has been particularly significant for the heaviest goods vehicles (Table 2.1). About 90% of the heavy goods fleet is diesel powered.

Emission Loads

2.18   *Nitrogen Oxides (NOx)*. Road transport as a whole is the largest source of NOx in the UK. It gave rise to about 1.3 million tonnes, 48% of the total, in 1989. Other major sources are power stations, producing 29% of the total, and manufacturing industry producing 10%[23] The emissions from diesel vehicles were estimated to be about 0.56 million tonnes, 21% of the overall total[24]: 19% from HGVs and PSVs, the remainder from vans, cars and taxis[21].

2.19   In projections of atmospheric emissions from road transport in the UK, prepared by the World Wide Fund for Nature (Figure 2.5), it is esti-mated that both total NOx emissions and the contribution from HGVs will fall until the middle of the next decade. Both will then start to rise again. By then, the effect of reductions made in the diesel emissions per vehicle in the UK by the implementation of the agreed limit values for 1993 and 1996 (paragraph 3.7 and Table 3.1) will have been outweighed by the forecast growth in vehicle numbers and distance travelled. The proportion of total road transport NOx coming from HGVs will grow, rising to nearly two thirds by the year 2020 as petrol driven cars fitted with three-way catalysts form an increasing proportion of the car fleet. Electricity generation is the other major source of NOx, accounting for 29% of the national total in 1989. As control of emissions from power stations is introduced this propor-tion will be expected to fall and the proportion from road transport to rise. The proportion of total NOx emitted from motor vehicles is higher in regions like London and the rest of south east England which have high traffic flows and lack heavy industry and large power stations. An emissions inventory carried out in 1983/4 established that traffic accounted for 70% of NOx emissions in Greater London and over 50% in large areas of southern Britain[25].

	1984	1989	1989 as a % of 1984
Vehicles (thousands)			
Diesel powered			
Goods (>1.5 tonnes unladen)	450	567	126
Public transport (includes taxis)	89	103	115
All methods of propulsion			
Heavy goods vehicles (tonnes gross weight)			
3.5 – 7.5	136	157	115
7.5 – 12	29	20	69
12 – 25	147	146	99
25 – 33	75	62	83
33 – 38	20	59	195
Total	407	444	109
Buses and coaches (with more than 8 seats)	67	73	108
All road vehicles (other than motor bikes)	18,288	21,931	119
Distance Travelled (billion kilometres)			
Goods (> 1.5 tonnes unladen)			
Rigid	15.55	19.49	125
Articulated			
3 & 4 axle	5.47	5.03	92
5 axle	1.3	5.11	390
Total	22.33	29.70	133
Buses and coaches	3.81	4.33	114
Freight Transported (billion tonne kilometres)			
All vehicles (>1.5 tonnes unladen)	97	132	137
Diesel Fuel Consumed (thousand tonnes)			
All diesel vehicles	6,755	10,118	150

Table 2.1   Growth in numbers and use of vehicles [2,22]

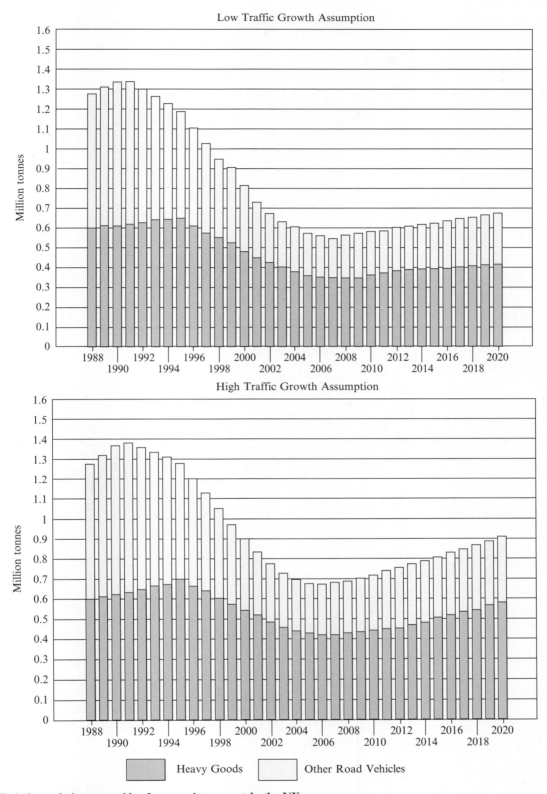

Figure 2.5 Emissions of nitrogen oxides from road transport in the UK.
Projections of nitrogen oxide emissions on the basis of 2 forecasts of demand growth issued by the Department of Transport in 1989: low forecast at top and high below. The projections take into account:

— implementation of the agreed amendments to EC Directive 88/77/EEC in 1993 and 1996:
— replacement and growth of the fleet of heavy good vehicles and passenger service vehicles (over 32 seats) as forecast by the Department of Transport.

Source: World Wide Fund for Nature[21].

2.20   The concentration of NOx, which is made up of nitrogen dioxide (NO_2) and nitric oxide (NO), in the atmosphere varies widely over the country. Concentrations in city centres and rural areas may differ by a factor of 10. A survey found annual mean NO_2 levels in central London of around 30–40 parts per billion (ppb) falling off to 20 ppb on the southern outskirts[25]. Annual mean levels in the remote countryside, by contrast, are of the order of 2 ppb. Variations in the emissions of NOx across Greater London are illustrated in Plate 4. More detailed surveys in London found annual mean levels of 80 ppb of NO_2 in the centre of the street falling off to 40 ppb at the back of the pavement. Variations in concentration over shorter periods can reach much higher levels and a record hourly mean level of 2200 ppb for NO_2 was recorded at one central London site in 1983 although typical maximum hourly concentrations are around 1000 ppb for both NO and NO_2. Although NOx levels in rural areas are relatively low, they have been rising sharply in recent years with a doubling reported at a number of sites between 1979 and 1987. In contrast there has been no clear evidence of an overall upward trend in NOx in the cities which have been monitored[25].

2.21   *Ozone.* At high altitudes ozone is produced from oxygen by the action of ultra-violet radiation. It forms the ozone layer which shields the surface of the earth from ultra-violet radiation. In the lower atmosphere ozone is formed by the action of sunlight on nitrogen dioxide in the presence of oxygen. NOx, in the presence of sunlight, catalyses the formation of ozone from hydrocarbons and oxygen. Ozone is a very reactive molecule and in the lower atmosphere has harmful effects on human health and vegetation. In this Report we are concerned only with the production of ozone in the lower atmosphere. The role of diesel emissions in the formation of ground level ozone is not straightforward. Diesel emissions are a source of NOx and hydrocarbons but modelling studies[26] have shown that the former is more important in the formation of ozone. Nitrogen dioxide, including that formed from nitric oxide emitted from diesel engines, promotes the formation of ozone whilst nitric oxide itself removes ozone. The mechanisms involved are described briefly in the box opposite.

2.22   As a result of the mechanisms described in the box, notably the role of hydrocarbons and of sunlight in the formation of ozone, concentrations of ozone tend to be higher in rural than in urban areas and higher during the summer than the winter. Base line levels in rural areas in the summer are in the range 20–40 ppb. Concentrations exceeding 60 ppb were recorded for 100–200 hours in 1989 in rural areas (compared with 75 hours in central London); exceeding 100 ppb for 10–50 hours, and exceeding 120 ppb for up to 10 hours[27]. Long term trends are obscured by the considerable year to year variability but it has been estimated that ozone concentrations have doubled in the last century[28].

2.23   *Particulate Matter.* The particulate matter emitted from diesel engines includes a high proportion of very fine particles, less than 1 micrometre in diameter. Diesel engines make the largest contribution to particulate elemental carbon material with annual emissions in 1986 of about 15,000 tonnes[29]. Their contribution to total UK emissions of black smoke has risen from 8% in 1970 to 32%, about 182,000 tonnes in 1989[24,30]. Aggregated particles are removed by dry deposition or washing out by rain, whilst fine particles provide condensation nuclei for rain drops. Most particles are deposited close to the source but some may remain in the air for days or weeks and may be transported over thousands of miles[31].

OZONE AND NOx IN THE LOWER ATMOSPHERE

At lower levels in the atmosphere ozone is produced by the reaction of oxygen molecules with atomic oxygen which is supplied by the breakdown of nitrogen dioxide in the presence of sunlight.

Formation of Ozone

$$NO_2 \xrightarrow{\text{sunlight}} NO + O$$

$$O_2 + O \longrightarrow O_3$$

The nitric oxide produced in this reaction would normally react rapidly with ozone to regenerate nitrogen dioxide in a reaction which is not dependent on sunlight and therefore continues through the night.

Removal of Ozone

$$NO + O_3 \longrightarrow NO_2 + O_2$$

Nitric oxide is also converted to NO_2 by oxygen and so may be oxidised without the consumption of ozone, allowing the concentration of ozone to increase.

With nitric oxide as well as nitrogen dioxide being present in diesel emissions, nitric oxide can build up to high concentrations around dense traffic. Because of ozone's reaction with this continuous supply of nitric oxide, concentrations of ozone near traffic are low. The processes that lead to the build up of ozone are not instantaneous. Over a period of days polluted air may be blown away from the source of emissions. The ozone concentration rises as vehicles no longer add nitric oxide in significant amounts and other reactions, involving oxidised products of hydrocarbons, continue to generate ozone. Hydrocarbons are derived from many natural and man-made sources; diesel engines are a relatively minor source.

Elevated ozone concentrations are also associated with the formation of raised concentrations of other pollutants, especially peroxyacetyl nitrate (a component of photochemical smog) and summer haze arising from the oxidation of sulphur dioxide and nitrogen oxides to produce aerosols of sulphates and nitrates.

2.24 *Sulphur Compounds.* The sulphate emitted as particulate matter is only about 2% of the total sulphur emitted from diesel engines; the remainder emerges as sulphur dioxide gas. Together these sulphur products contribute around 1% to the national load of sulphate deposition, the bulk of which is derived from stationary sources[23]. Concentrations of sulphur dioxide are high in London[24] and emissions from diesel engines may make an appreciable contribution to levels in city streets.

Air Quality

2.25 The European Community has established annual air quality limit values for nitrogen dioxide, sulphur dioxide and suspended particulates and guide values for nitrogen dioxide and sulphur dioxide. The World Health Organisation (WHO) has established guidelines for concentrations of nitrogen dioxide and ozone. Details are in the box below. The EC limit values for NOx for have been exceeded in some recent years at two of the three monitoring sites in London but not at other urban or rural sites in the UK. The guide values have been exceeded in most recent years in London and in some years in other cities[25]. The limit values for particulates have been exceeded in the neighbourhood of some large combustion plants. The WHO guidelines for ozone have been exceeded every year, at all rural monitoring sites with the exception of two remote sites, for the past five years. In 1984 the UK Government adopted a policy aim of reducing national NOx emissions, by the end of the 1990s, by 30% from the 1980 level. It now recognises that the growth in road traffic since then, and high growth for the rest of the decade, will make the achievement of this aim difficult[30]. Little systematic work appears to have been done on forecasting the impact on air quality of traffic growth and changes in vehicle emissions standards. This is an area that requires further study.

AIR QUALITY VALUES

Air quality limit values have been established by the European Community to protect human health. Limit values are specified in Directives for nitrogen dioxide[32] and for suspended particulates (and sulphur dioxide)[33]. The Directives are implemented in Great Britain by the Air Quality Standards Regulations 1989[34] made under the Control of Pollution Act 1974. The Regulations require the Secretary of State to ensure that the levels of pollutants are measured and kept below the specified values. The limit values are:

Nitrogen dioxide 200 microgrammes per m³ ($\mu g/m^3$) (equivalent to 105 ppb) for 98% of hourly mean concentrations in a calender year.

*Suspended particulate*s 250 $\mu g/m^3$ for 98% of daily mean concentrations in a calender year.

The Directives and Regulations also include guide values for nitrogen dioxide which are meant to act as targets for future air quality. They take into account a wider range of potential environmental impacts than those affecting human health. They are:

135 $\mu g/m^3$ for 98%, and 50 $\mu g/m^3$ for 50%, of hourly mean concentrations over a calender year.

Guidelines have been established by the World Health Organisation[35] for short term exposure to nitrogen dioxide and ozone. They allow a margin of protection below the minimum concentrations that are associated with adverse effects on the health of the general population. The guidelines are:

Nitrogen dioxide 400 $\mu g/m^3$ (209 ppb) for the hourly mean concentration and 150 $\mu g/m^3$ (78 ppb) for the 24 hourly mean concentration.

Ozone 150-200 $\mu g/m3$ (76–100 ppb) for the hourly mean concentration.

The Impact of Emissions

2.26 Diesel emissions are potentially harmful to the health and the quality of life of people and to the environment as a whole. Emissions are made close to ground level and may be prolonged and concentrated. In city streets large numbers of people and buildings are directly affected. Subsequently polluted air moves into the countryside adding to the emissions that are generated there. Crops and vegetation as well as the rural population can be affected.

Health Effects

2.27 *Carcinogenesis.* The Department of Health Committee on Carcinogens, the WHO International Agency for Research on Cancer (IARC) and the US National Institute of Occupational Safety and Health have evaluated diesel exhaust as a probable human carcinogen. Polyaromatic hydrocarbons

(PAHs) are found in diesel exhaust and include a number of known mutagens, chemicals which generate genetic mutations, and carcinogens such as benzo(a)pyrene. The PAHs are associated with the particulate matter rather than gaseous hydrocarbons in the exhaust. Diesel exhaust was considered by IARC to be probably carcinogenic to humans, whilst exhaust from petrol engines was considered to be possibly carcinogenic. The evidence is drawn from inhalation studies in animals and epidemiological surveys of people in occupational risk groups. Other relevant data have been obtained from mutagenicity assays.

Inhalation Studies. The fine particles emitted by diesel vehicles are able to penetrate into the deepest parts of the lung where they provide a mechanism for presenting mutagenic material to cells over a long period of time. Particulate concentrates introduced into the lungs of rats have been shown to cause a higher incidence of tumours than would be expected from an aerosol of benzo(a)pyrene. Several inhalation studies have been carried out by exposing experimental animals over long periods of time to emissions from diesel engines. Experiments with rats have shown a response related to dose in the production of tumours. When particulates were filtered from the gases tumours did not appear. The data refer mainly to light duty engines; few studies have examined heavy duty, direct injection engines and it is known that changes in engine type and operating conditions may lead to significant changes in components of the emissions, such as PAHs. The significance of these experiments has also been questioned because the massive concentrations of particulates appeared to overload the normal clearance mechanisms in the lungs. Nevertheless the IARC evaluation concluded that whole diesel engine exhaust is carcinogenic in experimental animals[36].

Epidemiology. IARC found that there was limited evidence for the carcinogenicity of diesel emissions in humans. Some studies among heavily exposed occupational groups, such as workers in diesel rail locomotive depots or bus garages, have shown an increased risk of lung cancer. Some studies have also indicated a risk of bladder cancer[36]. More recent work[37] on Swedish bus garage workers showed that risk increased with increasing cumulative exposure. It should be noted, however, that this study, like others in the field, was retrospective and did not control specifically for tobacco smoking which is the major influence on the development of lung cancer.

Mutagenicity assays. Investigations of known cancer inducing compounds have shown some correlation between mutagenicity and the degree of carcinogenicity. Diesel emissions were tested for the presence of mutagens with *in vitro* and *in vivo* assays in bacteria, yeast cells, cultured mammalian cells, insects and rats. The emissions induced chromosomal damage in cultured mammalian cells, insects and rats, and mutations and DNA damage in bacteria. In some bacterial assays it was found that the mutagenic effect of diesel exhaust was about ten times greater than the effect of the exhaust from petrol engines[38].

2.28 If diesel particulates are widely dispersed in the atmosphere they may be expected to contribute to the background burden of carcinogens on the population as a whole. Evidence for this is difficult to obtain as the magnitude of the risk demonstrated in heavily exposed occupational groups is about one tenth of the risk from tobacco smoking and cases of cancer that were due to diesel exhaust may be too few to distinguish. Coal smoke was another significant source of exposure to carcinogens for city dwellers in the past and has probably contributed a greater hazard than subsequent exposure to diesel exhaust. Lung cancer rates in the population of the UK as a whole have been falling in recent years, an effect that is probably due to the reduction in emissions of coal smoke and in tobacco smoking. There has been no indication from the trends in lung cancer that diesel exhaust, which has increased over the same period, has had a perceptible effect[39]. In the future it may still be difficult to demonstrate an association of lung cancer with diesel emissions as the incidence of the disease will be expected to continue to fall. This is due to the latency period in its development, which can be up to 40 years.

2.29 *Respiratory Conditions.* Nitrogen dioxide is a respiratory irritant that can, at sufficient concentrations, lead to temporary reductions in lung function in healthy people and more particularly in asthmatics[40]. There are indications that it may also reduce resistance to respiratory infections or increase sensitivity to allergens[41]. Concentrations which would be expected to produce such effects have occasionally been recorded in highly congested streets in London[42]. Ozone has somewhat analogous effects, with the risk of small reductions in lung function, especially when breathing deeply as in vigorous exercise. Concentrations within the range linked with such effects have occurred occasionally in the UK during periods of hot sunny weather, the highest values being outside major urban areas. Discomfort and eye and throat irritation are also experienced in heavy traffic. These are related to the emission of unburnt fuel and of gaseous hydrocarbons[43], derived for the most part from petrol engines.

2.30 *Future Developments.* Studies of the relationship between diesel emissions and cancer have so far established only a weak linkage. It may not be possible to demonstrate a relationship which can be quantified to the same extent as has been achieved with other, more potent carcinogens. Epidemiological studies are central to this issue but research is hampered by the exposure of subject groups to other likely causes of lung cancer, especially tobacco smoke. There remains scope, however, for more detailed mutagenic studies of individual compounds which may contribute a risk of cancer. Identified carcinogens are emitted by diesel engines but invariably appear as mixtures and very little is known of the interactions between the components of these mixtures. We recommend that further work should be done to identify the specific sources of carcinogenicity in diesel exhaust and the toxicological effects of defined fractions or combinations of components.

Environmental Effects

2.31 *Amenity.* The public perceive diesel emissions to be having a detrimental effect on the air that they breathe. The most obvious cause of complaint is black smoke. Other emissions, such as NOx and hydrocarbons, are less obvious but people suffer discomfort when breathing them and dislike their smell. Particulates cause a reduction in visibility; in Holland it has been estimated that diesel exhaust can be responsible for a 13% reduction in visibility in city streets[43].

2.32 *The Built Environment.* Since the levels of pollution from coal smoke have been reduced, diesel particulates have become the major cause of soiling of buildings[30]. They are not easily washed off by rain because of their small size and the adsorbed hydrocarbons which make them sticky. It has been estimated[44] that diesel particulates are three times as effective as coal smoke and six times as effective as particulates from petrol engines in soiling surfaces. The soiling of buildings is important not only because of the loss of visual amenity but also because of the structural damage that may result, the deposited material enhancing the effect of acidic gases and accelerating the erosion of the stone. These are illustrated in Plates 5a and 5b. As far as we are aware these effects of vehicle emissions in the UK have not been fully costed. The economic costs of cleaning buildings has been estimated. A survey of major companies showed that about £80 million was spent in the UK on cleaning in 1987. This figure was probably a considerable underestimate of the actual expenditure, as small companies were not surveyed, and it took no account of the potential expenditure on cleaning the many buildings which in fact are not cleaned[45].

2.33 *Vegetation.* Studies in the USA carried out by the National Crop Loss Assessment Network have shown that concentrations of ozone commonly recorded are high enough to reduce the yield of field-grown crops, such as soybean, corn and wheat[46]. Ozone is now considered to be the single most important pollutant affecting vegetation in the USA. Although the effects observed in the UK have not been as pronounced, damage to sensitive crops including peas, beans and clover has been reported after the occurrence of episodes of elevated ozone concentration[93]. Ozone has been also identified as a greenhouse gas[47]. NOx has been shown to affect plant growth under laboratory conditions but as yet there is little evidence that it has adverse effects on crop yield. Nitrogen deposition from the atmosphere, which is in part derived from NOx, can act as a plant nutrient on agricultural land but has been reported to have exceeded the critical loads set for more sensitive ecosystems[25]. NOx contributes to acid deposition but it is not yet clear to what extent deposition on rural sites is affected by the uptake of nitrate by plants[48]. Particulate matter is also deposited on the soil and PAHs are taken up into crops. Eating food contaminated in this way is thought to be the major source of exposure to PAHs by non-smokers[49].

Analysis of Emissions

2.34 The emissions from diesel engines that are of concern are not single substances but can be divided into three categories — NOx, particulate matter and hydrocarbons — each a mixture of substances. The individual substances have very different effects on the environment. Present policy takes no account of these differences. Before an engine type is approved for production its emissions are measured under controlled conditions of load and speed (paragraphs 3.22 and 3.23). The analytical techniques specified in the tests measure the total masses of NOx, particulate matter and hydrocarbons (and carbon monoxide) emitted but do not provide any measure of the separate components of each of these categories. Particulate matter, for example, is a mixture largely of elemental carbon, hydrocarbons (including PAHs), sulphate and water (paragraph 2.4). It is defined for the purposes of the test as material, of whatever chemical composition, which is retained on a specified filter under certain prescribed conditions of air flow, dilution, relative humidity etc[4]. We recommend that the UK Government should propose that the European Community develops more discriminating techniques for analysing emissions from heavy duty diesel vehicles. This should be done in liaison with other authorities, with a view to introducing internationally agreed protocols on the measurement of such emissions.

CHAPTER 3

NEW VEHICLES AND ENGINES

Introduction

3.1 There are two approaches to reducing emissions from the heavy duty fleet as a whole. One is to ensure that new vehicles meet progressively tighter standards. The other, which is considered in Chapter 4, is to control emissions from vehicles already in service. This Chapter considers the limit values which apply at present to new vehicles, those which have been agreed for the European Community for the rest of this decade and the procedures by which new engines are tested to ensure compliance with those values. An economic instrument to encourage the early replacement of vehicles by new, less polluting ones is recommended. These points apply chiefly to vehicles which are registered to travel on the road but, at the end of the Chapter, consideration is given to the controls to be applied to off-road uses of diesel engines.

3.2 The emission standards to be met by new vehicles — or more precisely by new engines — in this country are determined by Motor Vehicles (Construction and Use) Regulations issued under the Road Traffic Acts. The latest set of Regulations[50], issued in 1990 under the 1988 Act, contains limit values for gaseous emissions. The previous set[51], issued in 1986, contains a limit for emissions of smoke which is still in force. Most of this Chapter is concerned with gaseous and particulate emissions but there is first a brief comment on the control of smoke emissions.

Smoke

3.3 The 1986 Regulations carry forward (from earlier ones) the implementation of EC Directive 72/306/EEC[52], which was itself based indirectly on a British Standard set in 1964[53]. The engine is tested, under full load, at a series of different speeds. To comply with the Directive, the smoke opacity values recorded must lie below a defined curve. This was the only EC legislation on emissions from heavy duty diesel vehicles until the Directive described in the next section. Smoke is composed of solid particles and liquid droplets, both of which are covered by the new limit values on particulates (see paragraph 2.34 and Table 3.1). Consideration should therefore be given to the possibility of consolidating the limit on smoke emissions into the Directive on other emissions from heavy duty diesel vehicles, discussed below.

Other Emissions

3.4 The standards for control of gaseous and particulate emissions in this country also derive from European Community legislation. The 1990 Regulations referred to above implement a 'Directive relating to the measures to be taken against the emissions of gaseous pollutants from diesel engines for use in vehicles', EC Directive 88/77/EEC[4].

Limit Values for Emissions

The European Community

3.5 EC Directive 88/77/EEC defined limit values for emissions of carbon monoxide, hydrocarbons and nitrogen oxides. The values are given in Table 3.1. Separate values are set for 'type approval' (the testing of a single engine, representative of the type) and for 'conformity of production' (the sampling of engines taken from the production run, to ensure that standards of production are met). The values were not onerous for engines of modern design. Indeed, evidence submitted to us[8] indicates that, for buses at least, the limits for hydrocarbons and carbon monoxide were considerably less tight than could be achieved by an engine several years old and in an indifferent state of maintenance.

3.6   This Directive was perceived as the first step towards tight control of emissions. It required the European Commission to put forward a proposal for a further set of limit values, including one for particulate emissions. The text of the Commission's proposal for an amendment to the Directive[54] became available in July 1990, just as our study began. The Council of Ministers reached agreement to a common position[5] at a meeting of its Environment Council in March 1991. At the time our Report went to press the formal adoption of the amendment was expected shortly. It seems likely to be similar to the text agreed in March, which we refer to as "the agreed amendment" and which is taken as the basis of our comments.

3.7   The agreed amendment defines two further stages of tightening of limit values. The first is to be achieved by 1 July 1992 (for type approval) and 1 October 1993 (for conformity of production), the second by 1 October 1995 and 1 October 1996 respectively. It also introduces limit values for particulate matter. The values agreed are given in the table below. The three sets of values in the table are named and numbered in different ways by different commentators; we refer to them as "the EC Directive 88/77/EEC" and "Stages I and II of the agreed amendment". It was initially proposed that Stage II should be implemented in 1996 and 1997 and that the limit value for particulate emissions should not be determined until 1994 but the dates and values set out in the table are those now agreed. As an adjunct to the setting of a tight limit for particulates at Stage II, it has been agreed that the maximum permitted sulphur content of diesel fuel shall be reduced from 0.3% to 0.05% by 1996. The implications of this are considered in Chapter 5.

Table 3.1
Emission Limit Values Agreed* by the European Community

	Implementation dates		Limit Values in g/kW.h			
			Carbon Monoxide	Hydro-carbons	Nitrogen Oxides	Particulate Matter
Directive	TA	1. 7.88	11.2	2.4	14.4	no limit
88/77/EEC	COP	1.10.90	12.3	2.6	15.8	no limit
Amendment	TA	1. 7.92	4.5	1.1	8.0	0.61/0.36
Stage I	COP	1.10.93	4.9	1.23	9.0	0.68/0.40
Amendment	TA	1.10.95 ⎫	4.0	1.1	7.0	0.15
Stage II	COP	1.10.96 ⎭				

Notes: TA = Type Approval (see paragraph 3.5)
         COP = Conformity of Production
     * The status of the agreement to these limit values, at the time of going to press, is described in paragraph 3.6.

3.8   The composition of "particulate matter" in the above table is described in paragraphs 2.4 and 2.34. "Hydrocarbons" refers to unburnt gaseous hydrocarbons. Emissions are expressed in grams per kilowatt hour (g/kW.h) — a measure of the mass emitted per unit of the engine's average power output during the test (paragraph 3.22 and box) per hour of the test. Stage I of the agreed amendment contains two limits on particulates: one to be applied to engines whose maximum rated power is 85 kW or less and a lower one for engines of more than 85 kw. Thus the smaller engines are permitted to emit more grams per kilowatt, reflecting their generally simpler design and lower cost. There is no such distinction on the basis of engine size for Stage II of the agreed amendment. Nor is there a distinction between the values permitted for type approval and for conformity of production.

3.9 It is expected that there will be a further amendment to the Directive, imposing lower values, but it has been agreed that this will not come into force before 1 October 1999. A number of points were raised in evidence which related to the amendment then under discussion. Some related to the timing of the stages and the lead times for implementation by industry; these have now been agreed and are not discussed in this Report. Other points, however, are equally relevant to preparations for the further amendment at the end of the decade and are addressed in that context.

The United States of America

3.10 In the USA, national emission standards for diesel vehicles are set by the Environmental Protection Agency, though States may impose stricter standards. A rule issued by the Agency in 1985 set limit values (for conformity of production) for the same emissions as are now to be controlled in the European Community (EC), namely carbon monoxide, hydrocarbons, NOx and particulate matter. Initial values for the first three were set for 1987 models. Successively lower values for NOx, and for particulates, were set for model years 1988, 1991 and 1994. These are not precisely comparable with the EC values given in Table 3.1 because a different test method is employed (see paragraphs 3.22, 3.23 and 3.34) but roughly comparable values for NOx and particulate matter in g/kw.h are said[55, 56] to be:

	Nitrogen Oxides	Particulate Matter
1991	7.2	0.34
1994	7.2	0.13

Buses operating in urban areas were to be required to meet the 1994 value for particulates by 1991 but this requirement has recently been deferred to 1993[8].

3.11 It is striking that values similar to those to be met in the EC in 1996 are to be met in the USA in 1991 (for NOx) or 1994 (for particulate matter). We have received conflicting evidence as to whether there is any good reason for this difference in timing. One relevant factor appears to be that, in the USA, only the largest vehicles are diesel powered (the rest having petrol engines which are subject to different emission limits). Large diesel engines are better suited than small ones, both by physical size and cost, to receive tight control of particulates and NOx. Two other factors suggested to us[57] are that vehicles are required by EC regulation to meet tighter standards for smoke emissions than is the case in the USA and that the European market demands higher standards of fuel economy, the latter point being especially relevant to the control of NOx emissions (paragraph 2.13). All of these differences require additional engine development work to be done in Europe before comparable emission limit values may be achieved. In addition, it appears that the engine certification procedures in the EC may impose a delay in the implementation of agreed values; this is considered further in paragraphs 3.36-3.38.

Other Countries

3.12   Other countries are moving, at different rates, in the same direction as the EC and the USA. Japan, for instance, has established the following limit values (in g/kW.h) for NOx and particulate matter[50]:

	Nitrogen Oxides	Particulate Matter
1994	6.0	0.7
1999 (or earlier)	4.5	0.25

This may imply an approach slightly tighter on NOx control, but much less tight on particulates, than in the EC. The engine test is slightly different from that used in the EC (paragraph 3.22), however, in that it covers only low and medium engine speeds when formation of NOx tends to be lower, but that of particulates higher, than at high speed. Sweden has set a value of 9.0 g/kW.h for NOx and 0.4 g/kW.h for particulates, to be voluntary in 1991 and mandatory in 1994, whilst Austria and Switzerland are requiring 9.0 g/kW.h NOx and 0.7 g/kW.h particulates in 1991 and tighter limits thereafter[56]. These countries use the same kind of engine test as the EC.

The Basis for Limit Values

3.13 It appeared to us, and has been confirmed in oral evidence[59], that the European Community's limit values for heavy duty vehicle emissions have been set on the basis of what could reasonably be expected from technological development by the time the values were due to be implemented, bearing in mind the cost of the measures required. They appear also to have been influenced substantially by the US values, which they closely resemble and which in turn stem from an essentially technology-based approach. Such an approach represents, in a sense, the application of BATNEEC (best available techniques not entailing excessive cost) to the control of vehicle emissions.

3.14   We recommend that a more fundamental approach should be taken to setting emission limits for vehicles. A guide value for nitrogen dioxide has been identified by the EC as a target for air quality (see box on page 18). Guide values should also be established for other emissions including suspended particulates or their carbon and hydrocarbon components. Guide values should be set in the light of the best available data on the health effects and other impacts of the pollutants, including the costs of those impacts, and should be reviewed periodically.

3.15   Measures designed to move towards achievement of those values should then be devised, in the light of technical, financial and other considerations. These would include the setting of limit values for heavy duty and other vehicle emissions but would need also to address other factors such as:

the growth in vehicle numbers and use;

sources, other than vehicles, of the relevant pollutants;

variations over time and location of compliance with the guide values.

Comparison of actual air quality with the guide values would be used to judge the efficacy of the measures adopted and the priority to be accorded to the control of different sources. The development of such an approach will involve a substantial programme of work. The Government should commence the necessary programme.

3.16   Until that work has been done, we support the continuation of the present approach of aiming to reduce all emissions as far as is reasonably practicable. This is the most appropriate course of action in the light of present knowledge of the environmental impacts of the various pollutants. Developments in engine technology do not, however, present only one possible set of reductions in emissions. Rather, there may be a trade-off between the degree of control which can be achieved over different emissions, so that a choice between them is made either explicitly or implicitly.

3.17   The trade-off in the control, by engine design, of particulates and of nitrogen oxides, is described in paragraph 2.14 and is illustrated in Figures 2.3 and 2.4. The known effects which each has on human health and on the natural and built environments are described briefly in the same Chapter. The findings of the expert committees with respect to the carcinogenicity of diesel particulates justify a precautionary approach which seeks to reduce such emissions as far as is practicable. There is also a strong case for continuing to seek reductions in emissions of NOx. Road transport is the largest single source of NOx in the UK and its contribution is still growing in absolute terms and in relation to other sources. Heavy goods vehicles are forecast soon to contribute more than half of the road transport total (Figure 2.5). On the basis of present knowledge, we consider that future emission limits, including those to be implemented in the European Community at the end of the decade, should require tighter control of both NOx and particulate matter; they should not concentrate on one to the exclusion of the other. This policy will need to be reviewed in the light of the further work on the impacts of emissions which we recommend above. It may also be influenced by technical developments, as follows.

3.18   Decisions on the emission limits to be set by the European Community at the end of the decade will need to be taken during the mid-1990s. It may by then be clear that, by the end of the decade, it will be practicable to fit particulate traps (paragraph 2.9) to all heavy duty diesel vehicles. If not, however, and assuming that particulate emissions in urban areas continue to give rise to concern, consideration should be given to measures targeted on urban areas. These could include a much lower limit on particulate emissions from buses than from other heavy duty diesel vehicles, requiring buses to be fitted with particulate traps.

3.19   A similar approach should be taken to flow-through catalysts for controlling emissions of NOx (paragraph 2.10). By the mid-1990s it may be clear that, by the end of the decade, it will be practicable to fit them to all new heavy duty diesel vehicles. If not, however, consideration should be given to setting a much lower limit on NOx emissions from buses than from other heavy duty diesel vehicles, requiring buses to be fitted with such catalysts.

3.20   In paragraph 2.34 we recommend the development of more discriminating analysis and measurement of the components of diesel emissions. In the light of that work, consideration should be given to the setting of separate limit values for some of the various components.

Engine Test Cycles

3.21   There has been much debate recently about which engine test cycle should be used in the European Community. The matter has been settled for the present — the cycle laid down in the EC Directive 88/77/EEC will be retained for Stages I and II of the agreed amendment — but the European Commission is required to conduct a review of test procedures and to come forward with proposals. The arguments used are therefore of no less relevance in the context of the expected further stage of amendment at the end of the decade.

Steady State and Transient Cycles

3.22   The Directive and its agreed amendment define in considerable detail the engine test cycle, the sampling equipment and other aspects of the test procedure. These were developed for the UN Economic Commission for Europe (UNECE) during the early 1980s. The engine is tested under steady state conditions, that is, without acceleration or deceleration, at 13 different combinations of engine speed and load. The test is described in the box below.

EUROPEAN STEADY STATE ENGINE TEST CYCLE

The European steady state cycle is an adaptation of a cycle developed in the USA. The US cycle had 13 modes of engine speed and torque (load) and gave equal weight to each in calculating the total 'score'. The UNECE working group considered that this was not representative of European driving conditions, in which the engines of heavy duty vehicles are required, for much of the time, to provide high torque under conditions of medium engine speed. The emissions from each mode were therefore given an arithmetical weighting to reflect the relative importance of the conditions represented by that mode. The greatest weighting was given to two modes: one representing idling (near zero speed and torque) and one mid-speed and high torque. The engine is held at each mode for 6 minutes, the exhaust gas being sampled during the last 3 minutes of the mode. The sample is part of the exhaust flow taken from a dilution tunnel. At the end of each mode the engine is moved on to the next. The pollutant mass flow for each mode is measured and the total used, with the modal weighting factors and the sum of the net modal power outputs, to calculate the emissions. They are expressed in grams per kilowatt hour (g/kW.h), that is, grams of emission per unit power output during the test, per hour of the test.

3.23   The US Environmental Protection Agency (EPA) has gone on to develop a transient test cycle in which the engine is taken through continuously varying conditions of speed and load. This is described in the box below.

US TRANSIENT ENGINE TEST CYCLE

In a transient test cycle the engine is taken through continuously varying conditions of speed and torque (load) under precisely defined control. The conditions covered in the US transient cycle cover the full ranges but are concentrated on the upper end of the speed range and to some extent on the lower end of the torque range. The cycle is run twice, first from a 'cold' start (250°C) and again from a hot start, the results from the two runs being weighted 1:6 in the total score. Each run takes 20 minutes. Engine speed, torque and emissions are monitored every second. Sampling is carried out on a full-flow, constant-volume basis, that is an analysis of the whole mass of pollutant contained in a set volume of exhaust gas. The test includes what is known as a 'motoring' component, when the torque output for the engine is negative, representing conditions in which the vehicle is slowing and its momentum 'drives' the engine. Emissions are expressed in grams per brake horsepower hour, that is, grams of emission per unit of the engine's power output during the test, per hour of the test.

The Case for the US Transient Cycle

3.24   The UK Government proposed that the European Community should adopt the US transient test for Stage II of the amendment. That proposal was not accepted but it remains an option for any further amendment. The UK case([3]) may be put, briefly, as follows:-

(a)   In order to meet the limit values required under Stage II of the amendment, and any lower values which may subsequently be adopted, manufacturers will make increasing use of electronic control systems for various aspects of engine performance, notably the timing of fuel injection. Electronic controls may be set in such a way as to minimise emissions at the particular speed and load conditions specified in a test cycle, whilst optimising other qualities such as fuel consumption or power output (at the expense of higher emissions) under other conditions, resulting in emissions from an engine in actual use which are higher than those indicated by the test. This 'cycle beating' may be achieved easily in the case of a steady state cycle, and with a substantial distortion of emissions performance, but only with difficulty for a transient cycle and with much less distortion.

(b)   The use of turbochargers is already common on larger vehicles in Europe and will be used on many medium-sized vehicles in order to meet the Stage II emission limits. During normal vehicle operation the turbocharger speed lags behind changes in engine speed, resulting in momentary peaks of emissions during acceleration and deceleration. Such peaks would pass unrecorded by a steady state cycle but are recorded as part of a transient one.

(c)   In order to meet the Stage II limits for particulates some vehicles may need to be fitted with exhaust aftertreatment, probably in the form of a regenerative trap (paragraph 2.9 and box, also section 5.2 of Appendix 4). Regeneration occurs unpredictably during a steady state cycle and may therefore significantly distort the results, whereas it is more predictable under the strictly controlled conditions of a transient test.

(d)   The US transient test includes a cold start. This is important in view of the problems associated with smoke emissions from diesel vehicles starting from cold.

3.25   For all these reasons, the Government claims that the US transient test gives a more realistic assessment of the likely emissions performance of an engine under actual driving conditions than does a steady state test, as has been demonstrated in trials carried out by the US EPA([3]). The largest UK-based manufacturer of diesel engines for vehicles supports the Government view([8]). It adds the argument that the adoption of the US test, provided that it was accompanied by adoption of the US limit values, would achieve international harmonisation of standards and thus reduce the development effort required of those who wish to sell into both markets. A further argument in favour of the US test in particular, rather than any other possible transient test, was that it alone was already developed and could therefore be used for the Stage II limits. That argument does not hold for the further stage of limits, to be introduced at the end of the decade, as there would be time to develop other possibilities.

The Case Against the US Cycle

3.26   It is widely acknowledged that the European test is defective in the ways identified above. Nevertheless, several witnesses([19,57,60,61,62,63]) have maintained that the US transient test should not be adopted by the European Community. The principal points made are as follows:-

(a) Because of its emphasis on conditions of high engine speed at the full range of load (box with paragraph 3.23), the US cycle fails to represent adequately the mid-speed high load conditions which are emphasised in the European test and which are characteristic of European driving conditions, especially in urban areas. It therefore encourages the development of engines which might perform less well under European conditions with respect to emissions and fuel economy.

(b) The transient cycle requires test equipment which is much more expensive than for the steady state cycle. This applies particularly to the motoring component of the US cycle and the requirement for full-flow, constant volume sampling (box with paragraph 3.23). In addition, the US cycle takes considerably longer to run on the test facility and requires the employment of more highly skilled staff. The combined effect of these is to make the cost to the manufacturer of carrying out each test 3–5 times as much.

(c) It is by no means impossible to 'beat' the US cycle in the way described in paragraph 3.24a, as has been shown by US experience. The advantage of the US cycle in this respect is therefore not conclusive.

3.27 The UK Government has presented further evidence[3] which disputes the claim above about cost. The cost of the European test equipment has risen with the new requirement to record particulates, whilst automation has reduced the time taken to run a test on the US equipment and the manpower commitment. In capital terms, the difference has all but disappeared — US$ 1.45 million for the US test equipment against US$ 1.3 million for the European — and further reductions in cost of the US test equipment are expected[64]. The timing and manpower differences are similarly reduced.

Other Possible Test Cycles

3.28 At the request of the European Commission, industry is developing a modified cycle[8]. This comprises steady state modes but with timed transitions between them, so that the effects of aftertreatment may be predicted and allowed for. It has been argued, however[3] that, because this is not a truly transient cycle, it would not achieve the full benefits associated with such a cycle. Another proposal put to us[19,60] was that the steady state cycle should be retained but that a transient component should be added to it to create a composite cycle. This appears to have support within other EC Member States[16,63].

Driving Conditions

3.29 We have received a variety of views as to whether driving conditions are similar in America and in Europe. No objective work on this appears to have been carried out but both the UK and German Governments are now investigating driving conditions in their respective countries[3,16]. It has been argued[3,65] that this is of limited relevance to the choice of engine test cycle. The only purpose of such a test, it is said, is to provide a reasonable degree of assurance that an engine which performs satisfactorily under it will also perform satisfactorily under actual driving conditions. The US test covers a wide range of speed and load conditions and is therefore an adequate indicator in this respect.

Conclusion

3.30 The arguments against the continued use of the steady state test cycle, as set out in paragraph 3.24, appear conclusive. It will not provide a reliable indication of the likely emissions performance on the road of an engine which is fitted with electronic controls, turbocharging or exhaust aftertreatment, still less of one which is fitted with all three of these as will increasingly be the case.

3.31 Whether the steady state cycle should be replaced (rather than adapted) and, if so, by what and when, is less clear cut. Since any decision to change will not need to be taken until the mid-1990s, there is time for further debate within the European Community and more widely. We offer the following comments.

(a) There would be value in achieving wide international harmonisation in the equipment used to test vehicle engines, which give rise to pollution problems on a regional and global scale. The cycle of speed and load conditions over which the engine is taken, however, could be allowed to differ to represent the driving conditions of particular concern in each country or region.

(b) As noted in Chapters 1 and 2, the impact of diesel emissions is greatest in urban areas. It is important, therefore, that the European Community's test cycle should emphasise the representation of urban driving conditions. It is not yet clear to what extent these differ from other conditions but the work being done in Britain and in Germany should throw light on this. Whether any proposed test cycle adequately represents European urban driving conditions appears to be more significant than whether driving conditions overall in one country or region are different from those in another.

3.32 We conclude that the UK Government was right to propose adoption of the US transient test for Stage II of the amendment to the EC Directive 88/77/EEC. Recent developments indicating that many engines will be able to meet the Stage II limit for particulates without the use of exhaust aftertreatment weakens the case but does not eliminate it. The EC Council of Ministers has agreed, however, that the present test should be retained for Stage II. Because of the need for industry to have adequate time for engine development against a known test, we would not now argue that any change should be made for that Stage. We recommend however that, in considering the engine test to be used in the EC for the next stage of limit values, to take effect at the end of the decade, the UK Government should press the Community to develop one which uses equipment to the same specification as the US transient test but whose cycle emphasises the representation of European urban driving conditions. It should be defined well before the implementation date of the limit values to which it will apply.

3.33 We understand([59]) that the European Commission may consider the possibility of raising the issue of testing heavy duty diesel engines in the UNECE, seeking wider international agreement to a harmonised procedure. The UK Government should encourage it to do so. The OECD and UNEP may also have a role to play in seeking the widest possible international agreement.

The Link with Limit Values

3.34 One aspect of the test cycle which has received too little attention is its link with the question of what emission limit values are appropriate, especially for particulates. Because of its greater emphasis on conditions of medium speed and high load, the European steady state test is, in one respect, a stiffer challenge to the engine designer than is the US test. Because it ignores emissions generated under transient conditions, however, it is in another respect a less stiff challenge. Some recent data suggest that the latter consideration may be the more important, at least for engines having turbocharging and aftercooling[66]. It is therefore not possible to draw a direct comparison between the US and EC limit values, even after making the arithmetical conversion between the sets of units in which they are expressed.

3.35 If the European engine test cycle is to be changed to a transient one, the new emission limit values will have to take account of that. A more direct comparison with the US limit values will then be possible. It does not follow, however, that identical limit values should be adopted in Europe and in the USA. Quite apart from any difference of emphasis in the importance of reducing the various emissions, it may be appropriate to adopt different limit values to reflect any differences in the driving conditions represented by the European and US tests.

Family Certification

3.36 One other aspect of the US engine certification procedure has been drawn to our attention[8], namely the practice of 'family certification'. Engines are adapted in many ways, for instance with different components, for application in different kinds of vehicle. Each engine application involves a specific set of components. Family certification allows an engine manufacturer to divide the engine applications into families, using criteria laid down by the authorities, and to present for testing only one or two members of each family. The engine with the highest fuel consumption is normally tested. If the authority considers that another member of the family is likely to have higher emissions, it also is tested. The authority accepts that, if the engine(s) tested meet(s) the limit values for emissions, so too will other engines in that family. By reducing the number of engines requiring testing, family certification reduces the cost of certification to the manufacturer and to the authority. It also reduces the time taken for certification, thus allowing a shorter lead time between the setting of limit values and their implementation.

3.37 The European Commission and others[16,59] maintain that the European Community's procedures for engine certification already allow sufficient flexibility in this respect. In the EC 'worst case' engines are tested. This means that, in selecting the components to fit to an engine which is to be tested, the manufacturer must select components each of which, individually, is likely to lead to the highest emissions.

3.38 Both methods of testing raise complex issues of definition and selection. There appears to be a difference of view as to whether the present degree of flexibility in the EC is indeed sufficient. Our technical advice[66] is that, under the European system, many more engines need to be tested, resulting in greater development effort and delay in the implementation of adopted emission values. We understand that the UK Government favours introducing further flexibility into the EC engine test procedure, on the lines of family certification[3]. We recommend that it should press for this.

Costs and Incentives

The Cost of Compliance

3.39   It is not easy to estimate the cost of complying with the emission limit values described in this Chapter. There is continuing development of engines, not only in order to reduce emissions, and market pressures operate in ways which make the forecasting of costs uncertain. The Department of Transport has nevertheless attempted an estimate of the cost of complying with the Stage II values described in paragraph 3.7 and Table 3.1[3]. It is based on the assumption that all engines will have to have turbocharging and aftercooling (TCA), which only the larger ones have at present, and that present TCA engines will have to be replaced with ones of higher specification. This leads to the estimate that a 7.5 tonne lorry which at present costs £14,000 will have about £500 added to its price, whilst a 38 tonner costing £45,000 will have about £2,000 added. For the UK's heavy duty fleet as a whole, the additional capital cost of complying with the Stage II values is estimated to be a little under £800 million in total or £125 million per year. Total capital expenditure on new heavy duty vehicles was estimated to be of the order of £4.3 billion in 1989. In addition, those vehicles which already have a TCA engine would face a fuel penalty of about 4%: for the rest, the change to the inherently more economical TCA engine would result in no net fuel penalty as the result of complying with the new values. This would lead to an addition of about £13 million (net of tax) to the present national annual fuel bill for heavy duty vehicles of about £1,750 million. The Government's view[3] is that these costs are justified by the benefits to the environment. We support this view.

An Economic Instrument

3.40   As was noted in Chapter 1, we consider the control of vehicle emissions to be an area well suited to the application of economic instruments. EC Directive 88/77/EEC and the agreed amendment apply to engines, and hence for the most part to vehicles, manufactured after the respective implementation dates. They have no immediate effect on the existing fleet of vehicles. There would be benefit in speeding up the introduction of the cleaner vehicles in two ways:

> by encouraging manufacturers to produce engines which are able to meet the new emission values before the date on which they are required to do so; and

> by encouraging operators to purchase vehicles whose engines have been certificated to the new values, rather than continue to use their existing, more polluting vehicles; or to replace the engine in an existing vehicle with a new, less polluting one.

3.41   A practical form of incentive for the manufacture and the purchase of vehicles with less polluting engines, before the dates on which tighter emission limits became mandatory, would be differences in the rate of vehicle excise duty (VED). Such differences could also encourage the replacement of old vehicles with those meeting the current emission values. At present the rate of VED depends upon the size of the vehicle, measured either in gross weight (for a lorry) or seating capacity (for a bus or coach), and other factors such as the number of axles. We recommend that the rate of VED for a vehicle should also depend on whether its engine meets the emission values specified in successive stages of EC legislation. For new engines in new or existing vehicles, two such stages may be expected shortly to be adopted but will not be implemented for a few years, namely Stages I and II of the agreed amendment to EC Directive 88/77/EEC (paragraph 3.7). In considering the replacement of old vehicles with new ones it is relevant also to distinguish between the emission values of the original Directive 88/77/EEC and higher values. Thus VED in any particular size category could, when Stages I and II have been adopted, be levied at four different rates, as follows:

the highest rate, applying to vehicles whose engines do not meet the emission limit values of EC Directive 88/77/EEC;

another rate, for vehicles whose engines meet 88/77/EEC;

a lower rate, for vehicles able to meet Stage I;

the lowest rate, for any vehicle able to meet Stage II.

These rate categories could best be applied on the basis of the certificated performance of the engine. They should apply for the whole life of the vehicle, unless and until it is fitted with an engine which qualifies it for a different rate.

3.42 The agreed amendment to the Directive[4] includes an article providing that "The Member States may make provision for tax incentives [which]:

shall apply to all [vehicles] fitted with equipment allowing the European standards to be met in 1996 to be satisfied ahead of time;

shall cease upon the date for the compulsory entry into force of the emission values for new vehicles;

shall be of a value, for each type of vehicle, substantially lower than the actual cost of the equipment fitted to meet the values set and of its fitting on the vehicle."

It will be seen that this offers the opportunity for Member States to implement part of the scheme recommended above. We recommend that the UK Government should press for the Directive to be further amended to widen the scope for the application of economic instruments. In particular, the time for which incentives may be maintained should be extended and it should be permissible for them to be applied to vehicles with engines meeting the emission values specified in EC Directive 88/77/EEC or Stage I of the agreed amendment (to be implemented in 1993) as well as to vehicles meeting the Stage II values (1996).

3.43 The rates of VED could be set so that the effect on total revenue was neutral, taking account of the number of vehicles in each category. Those numbers will change as older, highly polluting vehicles are replaced by newer, clean ones; this would be reflected in periodic adjustment of the rates. The highest rate could be significantly higher than the present single rate for each size category and the lowest significantly lower. The periodic rate adjustment would raise still further the level of the highest rate, for the most polluting vehicles, whilst introducing a new low rate for those able to meet any further stage of EC legislation. This would create a continuing incentive for operators to move to the cleanest possible vehicle as soon as it came onto the market. The demand thus generated would create an enhanced market share for any manufacturer who could achieve the next level of reduction well ahead of the required implementation date.

3.44 There might be a case for increasing the total revenue from VED on heavy duty vehicles, in order to reflect the wider environmental costs they impose, in which case the various rates would be correspondingly higher. We have not considered the merits of this, however, and make no recommendation on it in this Report. If it were to be done, our recommended differential in rates on the basis of emissions performance would operate in the same way.

3.45   In order to gauge the possible efficacy of an incentive based on VED it is necessary to compare its possible size with the cost of replacing an old vehicle with a new one. For one of the largest vehicles, a 38 tonne articulated lorry, the present VED is about £3,000 per year. The first, second and third levels of VED described in paragraph 3.41 might be set at £3,500, £3,000 and £2,500 respectively. The operator of an old, relatively highly polluting lorry, who replaced it with a new one, would thus gain £500 per year if the replacement met the emission values of EC Directive 88/77/EEC and £1,000 per year if it met those of Stage I of the agreed amendment. The present cost of such a vehicle (meeting the values of 88/77/EEC) is around £45,000, with an additional cost of less than £1,000 for meeting the Stage I values. Vehicles need to be replaced periodically, about every 5–6 years for large articulated lorries, and have some residual sale value.

3.46   One cannot draw firm conclusions from these figures, which are in any case only one example. It appears, however, that the benefit in reduced VED could provide a significant incentive for purchasing a vehicle which met emissions values tighter than those in force, rather than the legal minimum, if one were available. It may also be sufficient to persuade an operator to replace a heavy goods vehicle at least a few years before it would otherwise have been replaced.

Buses

3.47   Buses, and other passenger service vehicles such as coaches, pay a much lower level of VED than do heavy goods vehicles: only £450 for the largest vehicles. The degree of incentive which could be created by establishing multiple rates of VED would therefore be limited. It would be possible, in principle, to increase the rates of VED paid by passenger service vehicles. We do not recommend such a step because we do not wish to discourage the use of buses by increasing their costs but the matter merits further consideration.

3.48   For reasons which are explained in Chapter 1, we attach especial importance to the control of emissions from buses. In paragraph 4.33 we recommend that a grant should be offered for the retrofitting of a particulate trap to a bus. A similar grant should be offered for the fitting of a trap to a new bus, unless and until the use of such a trap is required by the setting of tight new limits on particulate emissions from buses (paragraph 3.18). In paragraph 4.38 we recommend that, subject to certain conditions, a grant should be paid for the replacement of the engine in a bus. A grant should also be paid towards the cost of a new engine in a new bus, meeting the emission values specified in one of successive stages of EC legislation, purchased before the relevant stage was implemented and the values became mandatory.

Off-Road Uses of Diesel Engines

3.49   Diesel engines are used off the road in vehicles and machinery, primarily for manufacturing (forklift trucks, generators, etc), construction and mining (bulldozers, mobile compressors, etc) and agriculture (tractors, combine harvesters, etc). The uses of diesel engines for other modes of transport, such as trains and ships, are not classed as 'off-road' uses and are not considered here.

3.50   The number of diesel engines in off-road uses is not recorded but it is certainly very large. They vary widely in size and in the length of time for which they run. A helpful indication of the scale of their use, relative to on-road use, may be obtained from the figures for fuel consumption. On the basis of figures published by the Department of Energy[22] we estimate that, in the UK in 1989, off-road uses as a whole accounted for about 22% of all diesel fuel consumption in engines (that is, excluding fuel for heating), as compared with about 66% for on-road uses. The remainder is made up by marine and rail transport. The contribution which off-road uses make to emissions is likely to be at least as large in proportion to on-road uses as is their fuel consumption, that is, at least a third of on-road diesel engine emissions.

3.51   Off-road vehicles and machinery are generally fitted with derivatives of on-road vehicle engines but these are often set up with simpler and less costly fuel injection systems, with reduced power output, resulting in larger emissions per unit of power[65]. In high concentrations they may give rise to significant local damage to amenity, as in the case of an open-cast mining site close to housing which was drawn to our attention[67].

3.52   Many of the uses of diesel engines in enclosed spaces, such as forklift trucks in warehouses and locomotives in mines, are subject to regulation by the Health and Safety Executive. Agricultural tractors are subject to an EC Directive[68] which limits the emission of smoke. The limit value is similar to that applied to road vehicles (as described in paragraph 3.3) but is less strict in that the engine is tested at only 80% of full load. The UK Government has proposed that the standard should be tightened and that the Directive be amended to include limits for other emissions[3]. We commend this approach.

3.53   Industrial tractors, works trucks and engineering plant are at present subject to no emissions control. Vehicles first used on or after 1 April 1993, however, will have to comply with the emission limits of an UNECE test, which are similar to those of EC Directive 88/77/EEC (paragraph 3.5) but less demanding. Many other kinds of off-road use of diesel engines are at present subject to no emissions control. Responsibility for off-road uses has been divided between several Government Departments, with no one of them taking the lead. We consider this to be an unsatisfactory situation.

3.54   There is, however, an internationally recognised non-mandatory Standard[69] for emissions from off-road uses of diesel engines, issued by the International Standards Organisation (ISO). This specifies test methods and emission values for different categories of engine. The tests are similar to the European steady state test but have been adapted to reflect the different operating patterns of different categories of off-road use. ISO is preparing an amendment to the Standard, setting limit values whose attainment would require technology similar to that required for the attainment of Stage I of the agreed amendment to EC Directive 88/77/EEC for on-road vehicles[3].

3.55   It appears that the Department of Transport, with encouragement from UK industry, has recognised the need for action to control the emissions from off-road uses of diesel engines. Together they are developing a set of proposals[3]. The main one is that an EC Directive should be prepared, requiring new off-road diesel engines to comply with certain limit values, using test methods such as those in the proposed new ISO Standard. The current smoke test would be replaced with a simplified version of the current US transient test for smoke. The Californian authorities, which have previously influenced US policy in this area, are also considering the merits of such a test. There is the possibility, therefore, of wide international agreement on the test methods and limit values for emissions from off-road uses of diesel engines. The UK Government has introduced the topic at the European Commission's Motor Vehicle Emissions Group, recognising the close connection between on and off-road engines noted above. We welcome the Government's initiative in proposing that an EC Directive be prepared and recommend its speedy development.

Plate 5(a) A soiled building on a city street, four years after the building had been cleaned. The effects of vehicle emissions can be seen by comparison of the soiled stonework close to the ground and the cleaner walls of the upper stories.

Photograph by courtesy of Dr T Mansfield

Plate 5(b) Erosion of the limestone structure of a building in Cheltenham, Gloucestershire. Vehicle emissions contribute to acid rain, which corrodes limestone. Diesel particulates enhance the effects of acid rain on stone surfaces.

Photograph by courtesy of Environmental Picture Library/Mike Jackson

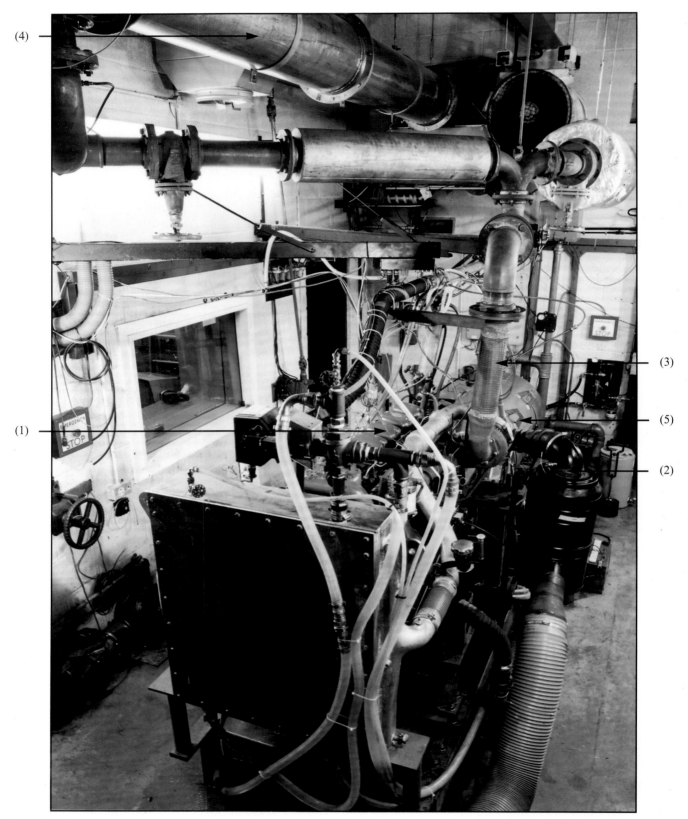

Plate 6   An engine under test on a transient test rig. The test installation includes control and analytical equipment in a separate room (not shown). A heavy duty diesel engine (1) takes in air (2) and passes its emissions through an exhaust pipe (3) into a dilution tunnel (4). Raw exhaust is mixed with air in the dilution tunnel to allow the physical processes of condensation and agglomeration that lead to the production of particulates. Samples are taken from the end of the tunnel. The dynamometer (5) absorbs the power output from the engine.

Photograph by courtesy of Ricardo Consulting Engineers Ltd

Plate 7   A particulate trap installed on a vehicle. The casing contains a ceramic monolith filter which traps particlate material. The particulate material is burnt off by an electric heater fitted at the base of the unit. Exhaust gases leave from the pipe at the top of the unit.

Photograph by courtesy of Volvo Trucks Ltd

Plate 8   A heavy dump truck (63 tonnes) working off the road.

Photograph by courtesy of VME Construction Equipment GB Ltd

(a)

(b)

Plate 9   Buses fuelled by compressed natural gas operate in Sweden. Fuel is stored under pressure in large carbon fibre cylinders which are mounted either (a) on the chassis or (b) on the roof.

Photograph by courtesy of Ecotraffic AB

CHAPTER 4

VEHICLES IN SERVICE

Introduction

4.1 Modern diesel engines are designed to give much tighter control of emissions than those manufactured ten years ago (paragraph 2.5), many of which are still in use. The emission values required by the agreed stages of amendment to EC Directive 88/77/EEC, described in the previous Chapter, are tighter still. Emission levels may, however, increase with prolonged use. Vehicles in service may therefore be expected to have higher levels of most emissions, per unit of power, than new vehicles, both because they were designed to less stringent standards and because they no longer perform to the standard to which they were designed. This Chapter considers the standards which should apply to vehicles in service, how they may be tested and enforced and the scope for achieving tighter control of emissions by modification or replacement of the engine and exhaust systems.

Emission Standards and Test Methods

4.2 A diesel engine is able to perform relatively efficiently, that is, to sustain its rated power output, for long periods with very limited maintenance. This is one of the attractions of this type of engine for the operator. Emissions may, however, increase for a number of reasons, including deposits on the fuel injector nozzles leading to sub-optimal fuel spray patterns and wear of the cylinders or piston rings leading to increased consumption of lubricant. Emissions of particulates and unburnt hydrocarbons are likely to increase. NOx emissions are likely to remain more constant or even to decline as the efficiency of combustion falls.

4.3 There are at present no regulatory requirements on emissions from a heavy duty vehicle in service, with the exception of smoke as described in paragraph 4.9. We consider that requirements should be imposed relating to the same range of emissions as was specified in legislation when the vehicle or engine was new. Several witnesses ([70,71,72]) suggested that there should be further legislation imposing standards for emissions from vehicles in service.

4.4 There is a severe practical problem to be overcome in implementing such an approach, however. In order to obtain a reliable indication of emissions from a diesel vehicle on the road it is necessary to measure them with the engine operating under a controlled load. The equipment required for this includes a chassis dynamometer, that is, a set of rollers which simulate the movement of the road in relation to the vehicle and which are able to apply a controlled load to the wheels and hence to the engine. For a heavy duty vehicle such equipment is very expensive — of the order of £1 million each including the analytical equipment — and could not reasonably be provided on the scale required to test all vehicles annually. It is therefore necessary to seek other ways of stating the requirement for vehicles in service to perform satisfactorily in respect of emissions control. Three such ways are described below.

4.5 First, in order to achieve sustained control of emissions over a prolonged period it is desirable that the engine should be designed to be robust against deterioration. It has been suggested([71]) that certain features of the modern engine, such as optimised spray patterns and narrow piston rings, which have been developed in order to reduce emissions, may in principle be more liable to deterioration leading to loss of emissions control. It appears that this need not happen, however, and that adequately robust design is possible. The US authorities require that engines are designed not to deteriorate in emissions performance by more than a specified factor over prolonged periods of operation. We recommend that such an approach should be considered by the UK Government for incorporation into the European Community's engine certification procedures.

4.6 Another possible approach is to require that engines are maintained appropriately. To some extent this will be a consequence of the annual test procedure, and the proposed additional test described in the following section, since failure of the test might indicate a lack of proper maintenance. It has been suggested[73], however, that operators should be under a separate and explicit obligation to carry out proper maintenance, such as periodic cleaning or replacement of fuel injectors. A check that such maintenance had been carried out could form part of the annual test; the criterion might be production of certificates or invoices from accredited garages or depots, or validated entries in a log book of maintenance. There are practical problems in the way of this, including the difficulty of defining the maintenance steps which are necessary in different patterns of vehicle operation and the consequent likelihood of requiring action to be taken which is not in fact needed. Nevertheless, we recommend that the practicability of requiring that appropriate maintenance be carried out on heavy duty diesel engines should be considered.

4.7 A third approach, and one which appears to us to have particular merit, is that an operator should be required to keep an engine in good condition. There is a limited number of parameters which, if checked and found to be satisfactory, indicate that emissions are likely to be adequately controlled. Techniques for testing those parameters, known as diagnostic techniques, already exist and are being developed rapidly as electronic monitoring and control systems become more widely applied to engines (see paragraph 4.12 and box). It will shortly be possible to require that an engine should be kept in good condition as defined by an appropriate set of parameters. The Government should develop this approach as part of its own legislative control of emissions from heavy duty diesel vehicles in service and should propose its introduction throughout the European Community. The implications of this for test procedures are addressed in paragraphs 4.12–4.14.

Training

4.8 An important factor contributing to the sustaining of an engine in good condition is the care with which it is maintained. This depends in part upon the expertise of the maintenance staff. Diesel engines are becoming increasingly complex and finely adjusted. The use of diagnostic techniques by the operator for self-monitoring will demand a high level of expertise. It is possible that not all operators will have staff of the necessary expertise either to maintain their vehicle engines in peak condition or to make full use of the opportunities for advanced techniques of diagnosis. We recommend that the Government, in liaison with the relevant trade associations, should ensure that the standards of training and qualification for those maintaining heavy duty vehicles are reviewed and should initiate any action needed to ensure that the necessary standards are achieved by all operators.

Enforcement of Standards

Annual Tests

4.9   It is essential that the standards set for heavy duty vehicles be monitored and enforced. At present the only check is on smoke, reflecting the fact that only smoke is at present the subject of a standard for in-service vehicles. It forms part of the annual mechanical test which, for heavy duty vehicles, is carried out by the Vehicle Inspectorate at its testing stations. The technique used is the free acceleration smoke (FAS) test under which the engine, in neutral gear, is repeatedly accelerated from idle to maximum speed and allowed to fall again to idle. Smoke emissions are observed during acceleration. Their density is assessed visually, by trained Vehicle Inspectors, against a requirement that they should be "acceptable". The meaning to be given to this term was defined by a series of tests carried out in 1964 whereby a representative panel was asked to judge the acceptability or otherwise of smoke from lorries; the opacity of smoke deemed acceptable by half the panel was taken as the criterion[3]. The failure rate for this test is about 1%. The Vehicle Inspectorate is conducting trials with an opacity meter which would allow the test to be put onto an instrumented basis, with the aim of introducing the equipment into testing stations in the summer of 1992. This was recommended in the Royal Commission's Tenth Report[1] and we are pleased that it is now being put into effect. We understand that it forms part of a wider review of vehicle testing methods.

4.10   The Government intends to amend the Regulations which set the criterion for smoke emissions in this test. Instead of requiring that smoke emissions should be "acceptable", vehicles in service will not be permitted to emit "any avoidable smoke or avoidable visible vapour". The latter phrase is at present used in the Regulations as the criterion for new vehicles and is interpreted by the Government as being a more severe criterion than the present one for vehicles in service. No immediate change is proposed to the standard required in the test because it is not considered feasible to judge, by eye, the presence or absence of smoke at this lower level. When instruments become available next year, however, the test will be made more severe to fit the new criterion. We welcome this development and urge the Government to introduce the change as soon as possible.

4.11   The German Government is also introducing an instrumented smoke test for heavy duty vehicles[16]. The preferred (though not the only) method will involve running the vehicle on free rollers, against a load applied by its own brake. This overcomes, at least in part, the principal limitation of the FAS test, giving a more reliable indication of the emissions likely to be generated on the road. The UK Government should consider introducing such a refinement to the test procedure in this country.

4.12   A test of visible smoke, though valuable, is not sufficient as an indication of emissions. Invisible particulates and gaseous emissions are also of concern and poor performance in controlling them will not necessarily result in emissions of visible smoke[66]. Because of the practical problems involved in measuring directly the full range of emissions from heavy duty vehicles, it appears to us that the most fruitful approach may be that of diagnostic testing, as described in the box overleaf. Diagnostic testing can be applied easily only to engines of modern design, for which detailed reference data are available. It would be possible to obtain reference data on engines of slightly older design, but which are still in production, but probably not on engines which are no longer in production. We recommend that development work

DIAGNOSTIC TECHNIQUES FOR EMISSIONS CONTROL

Diagnostic techniques rest on the premise that, if an engine is in essentially the same condition as it was when built, it may be assumed to be conforming to its certificated level of emissions control. Its condition may be judged by certain parameters which are able to be measured and interpreted by electronic equipment. Diagnostic techniques have been developed for use in cars, especially in the USA, and are beginning to be applied to heavy duty vehicles.

The parameters which could be relevant include:

— engine speed;

— fuel injection timing;

— cylinder compression;

— power;

— electrical equipment; and

— pressure and temperature characteristics of turbocharger, cylinder and exhaust system.

The equipment available to measure these is non-intrusive — that is, it depends upon measurements which can be taken by fitting sensors to the outside of the engine. Injection timing, for instance, may be measured by clamping a pressure transducer to the high pressure fuel pipe and comparing the timing of the pulse of fuel with the rotation of the engine as indicated by a mark on the flywheel. The position of that mark, as compared with a reference mark on the engine block, may be determined either by use of a stroboscopic lamp or fitting a sensor to the engine; some diesel car manufacturers are already fitting such sensors as standard. Further developments in diagnostic testing will require sensors to be fitted onto or into the engine in areas such as the manifold and filters.

At present the fitting of the sensors and the interpretation of results of diagnostic testing requires skilled staff and the availability of additional numerical analysis. Developments in measurement techniques, and in the programming of the workshop equipment, should reduce these requirements.

The judgment of engine condition depends upon a comparison with the condition when new. By carrying out a diagnostic test on a brand new engine of the same design it is possible to build up a set of reference data. The test would be more simple, accurate and reliable, however, if the relevant parameters were measured on each engine, or on a sample of engines, after production, using equipment similar to that for in-service testing, and recorded in machine-readable form on the engine.

Another development is of on-board diagnostic facilities for the driver, alerting him to a wide range of aspects of the condition of the vehicle. These could include the parameters relevant to the emissions performance of the engine and of any exhaust aftertreatment system.

be carried out to determine the range of heavy duty engines to which diagnostic testing may be applied and to extend that range as far as is practicable. A test of engine condition should be added to the annual test for heavy duty vehicles in this country.

4.13 Engine diagnostic techniques are at only an early stage of development. Current developments are driven as much by the goal of optimal performance of the vehicle over its whole life as by the need to control emissions. Research should be carried out into the application of advanced engine diagnostic techniques to emissions control. The Government should sponsor such research if necessary. It should also consider the feasibility of requiring the incorporation of the necessary sensors and other equipment into new heavy duty engines, so that more comprehensive and informative testing may be carried out in the future.

4.14   It is important that all EC Member States should adopt adequate procedures for testing emissions from heavy duty vehicles. Many vehicles registered in other EC Member States travel on British roads and their number seems likely to increase. We were therefore pleased to learn[59] that the European Commission is preparing proposals for a Directive which would require annual testing to be carried out. The proposals are not yet firm but we understand that they are likely to involve only a test of smoke, as an indication of whether the engine is in an adequate state of maintenance, and that the test will be an instrumented FAS test as described in paragraph 4.9. These limitations, which may be necessary in the short term in order to win approval by Member States, should be recognised as falling far short of the ideal. The UK Government should press for the introduction, throughout the European Community, of diagnostic engine testing and possibly for the introduction of a 'roller' test for smoke as described in paragraph 4.11.

4.15   Though we have advocated diagnostic testing because it appears likely to be effective, it and the other approaches described in the previous section have another advantage over the smoke test. The latter tests engines against a single standard which all must meet. Such a standard is necessarily set at a level which almost all vehicles are able to meet if properly maintained. Some vehicles would therefore be within the limit, even if poorly maintained, yet capable of performing better still if properly maintained. In order to reduce the total levels of emissions as rapidly as possible, and to keep them down in the face of rising distance travelled and engine power, it is important that each vehicle should be required to perform as nearly as possible to its peak of emissions control. The requirement for an engine to be kept in good condition achieves this objective.

Other Checks

4.16   The annual test of vehicles, though essential, is insufficient as a means of enforcement of emissions control. An engine may be prepared for it but then allowed to spend most of the year operating in a manner which results in higher emissions. This may be done deliberately, in order to maximise power output, or by default. The annual test must therefore be supplemented by spot checks.

4.17   A limited number of spot checks are carried out under which vehicles are stopped and tested for safety and for smoke emissions. Tests are carried out by staff of the Vehicle Inspectorate with assistance from a Police Constable. Passenger service vehicles are not subjected to such checks whilst in use because the delay to the vehicle is considered to be an unreasonable imposition on the passengers[3]. Inspectors do, however, check vehicles for excessive smoke at places such as coach parks and bus stations and when they make periodic checks of fleet vehicles at operators' premises. We recommend that the Vehicle Inspectorate should carry out more such spot checks; we welcome the Government's commitment in the 1990 environment White Paper[9] to do so. The checks should incorporate the new instrumented measure of smoke and diagnostic techniques to indicate performance in controlling other emissions. Portable diagnostic equipment, which could be used in such checks, is already available[8] and is becoming increasingly versatile and reliable.

4.18   The importance of strict checks on the emissions from buses is heightened by the commercial pressures on operators resulting from deregulation of bus services and by the impact which buses have on the urban environment. We welcome another statement in the White Paper that the Government will place considerable emphasis on the need to maintain good emissions performance when licensing heavy goods vehicle operators and will be reminding operators of this need. Similar emphasis should be placed on emissions performance when licensing operators of passenger service vehicles.

4.19   The Police have powers to stop and check a vehicle emitting dense smoke, it being an offence to emit smoke in such a manner that other road users are endangered[51]. The number of prosecutions under these powers is relatively small: 647 in 1986, the last year for which national figures have been compiled. This is hardly surprising in view of the quantity and density of smoke which would appear to be necessary to constitute such an offence. In view of the expected continuing decline in the number of heavily smoking vehicles, referred to below, it seems likely that the number of prosecutions has fallen since 1986 and will fall still further. Nevertheless, this remains a useful sanction against the worst cases of pollution.

4.20   The Vehicle Inspectorate conducts an annual (previously bi-annual) survey of smoke emissions from heavy duty vehicles. This is carried out at a number of sites — the same each time since 1985 — and involves observing the smoke emissions from passing vehicles. The sites selected are out of urban areas, on fairly steep slopes, and the vehicles observed are those driving uphill. This means that vehicles are observed under conditions of high load, when smoke emissions are likely to be relatively high. It has the incidental effect that buses are largely excluded from the survey. The Inspectorate should seek to extend the scope of this survey by investigating new techniques for the remote sensing of a range of diesel emissions. These are being developed for the measurement of carbon monoxide from petrol engined vehicles[74] and may be adaptable to the principal emissions from heavy duty diesel vehicles.

4.21   The proportion of vehicles emitting visible smoke, as recorded in the survey, has fallen over recent years. In spring 1984 it was 13%, in spring 1987 7.7% and in spring 1990 2.8%[3]. The number of heavy duty vehicles increased by about 6% over that period, so there has been a sharp drop in the absolute numbers smoking as well as the proportion. The technical advice which we have received[66] suggests that the proportion and number is likely to continue to decline as older vehicles are replaced by those which meet the emission values specified in the present EC Directive 88/77/EEC (even though that Directive imposes no limit on particulate emissions) and then by those meeting the values agreed for 1993 and 1996 which include increasingly severe limits on particulate emissions (Table 3.1). Though not all particulate emissions give rise to visible smoke, there is a correlation between the two. Engines designed to the new values will, even if poorly maintained, not be likely to emit visible smoke.

4.22   The National Society for Clean Air and Environmental Protection (NSCA) has proposed[70] that there should be a more comprehensive system of spot checks on emissions, including visible smoke, from vehicles. It favours giving a leading role in this to local authority environmental health officers (EHOs), or else to the Vehicle Inspectorate or the Police. The NSCA has also advocated the development by EHOs of public campaigns to spot smoky vehicles. Such a programme has been run in the State of Victoria,

Australia, by the State Environmental Protection Authority, as described in the box.

VEHICLE SPOTTER PROGRAMMES IN VICTORIA, AUSTRALIA

The Environmental Protection Authority of Victoria, Australia, has developed a wide-ranging programme against vehicle smoke, noise or tampering with emission controls. Vehicles can be required to be presented at officially designated test sites. Checks are made of whether dealers in second hand vehicles are selling vehicles which fail to comply with emissions regulations. Enforcement is carried out by EPA staff with the cooperation of the police, the vehicle registration authority and the Ministry of Consumer Affairs. Action taken against offenders ranges from a simple courtesy letter through on-the-spot fines to prosecution.

As part of this programme, the EPA aims to spot smoky or noisy vehicles in use. This is based on a simple visual assessment of nuisance by trained staff. EPA staff and the police report vehicles which they consider to be excessively smoky or noisy, for instance any which emit visible smoke for more than ten seconds. The EPA traces the owner of such a vehicle through the registration computer and writes to them, advising them that the vehicle has been assessed as a nuisance and requiring them to have it checked at an approved testing station within a month. The satisfactory reply rate is about 95%.

As an adjunct to the regular spotting programme, the EPA conducts an annual 'Smoky Vehicle Public Awareness Programme'. Members of the public are encouraged to telephone a special number to report vehicles which they consider to be emitting excessive smoke. The EPA writes to the owner, asking them to check their vehicle. No further action is taken. The first such programme resulted in 2,500 calls in two months and calls continued long after the end of the campaign, indicating a high level of public concern.

4.23   In this country, variants on the two Victoria spotter programmes have been tried by a few local authorities, some involving members of the public. Early results from a trial in Derby[75] involving the public indicate that:

> the initial public response was substantial;

> the response declined and needed renewal by repeated advertisement;

> the majority of vehicles reported were local buses; and

> the bus companies concerned readily agreed to take remedial action.

It would be wrong to draw definitive conclusions from a few months of one trial. If these results were to be sustained, however, and repeated elsewhere, they would suggest that spotter campaigns involving members of the public have some public appeal but that it might be possible to take equally effective action against smoky vehicles without the involvement of the public. We consider that spotter programmes should be developed, with or without public involvement. More local authorities should consider introducing them and the Vehicle Inspectorate, the Police and the Driver and Vehicle Licensing Centre should cooperate fully with them.

4.24   We also commend the Government's recent initiatives in encouraging members of the public to report smoky vehicles direct to the Vehicle Inspectorate, including publicising the telephone numbers which may be used. Permanent arrangements should be made to ensure that the public is aware of how to report smoky vehicles to the Inspectorate.

4.25   It appears likely that the number of vehicles emitting smoke which is visible to the naked eye will continue to decrease, as noted in paragraph 4.21. More sensitive and versatile techniques will therefore be required to identify vehicles whose emissions have deteriorated in service, including remote sensing devices and diagnostic testing. We consider this to be a task most suited to the Vehicle Inspectorate, which, we have recommended, should develop such techniques and extend its programme of spot checks. This supersedes the recommendation in the Royal Commission's Tenth Report[1], that powers should be given to local authorities to take proceedings directly against the operators of vehicles emitting excessive smoke. If such a power had been provided at the time of the recommendation, it would now be serving a useful purpose; it appears, however, to be a power whose value would diminish over the next decade.

Retrofit, Rebuild and Replacement of Engines

4.26   The discussion above has focused on means of ensuring that the emissions from a vehicle in service are controlled as well as can reasonably be expected, bearing in mind the standards to which the engine was designed. It may, however, be possible to improve on those standards, in one or more of four ways:

by retrofitting equipment to the engine and exhaust system;

by rebuilding the engine;

by replacing the engine with another of more advanced design; or

by changing to a fuel which burns more 'cleanly'.

The first three of these are considered in turn below. The fourth, which may also involve one of the others, is considered in Chapter 5. The importance of such approaches depends upon the operational life of the vehicle; this is considered first.

Operational Life

4.27   About 63% of heavy goods vehicles (HGVs) are still on the road after 8 years of use[21]. The equivalent figure for cars (petrol and diesel combined) is 92%. About 18% of HGVs are still on the road after 12 years, compared with 42% of cars. Within the heavy duty fleet there appear to be wide variations in operational life: many of the largest vehicles travel long distances, cover a high mileage (100,000 miles per year is not uncommon) and require replacement relatively soon, whilst smaller vehicles travel more locally and may continue for many years.

4.28   No directly comparable data on passenger service vehicles are available. It is clear from observation, however, and the Department of Transport has confirmed, that a high proportion of 'traditional' double and single decker buses survive for more than 12 years, some for much longer. (Many long distance coaches, by contrast, follow a pattern much more similar to that of the largest goods vehicles.) Again there are operational reasons for this, such as the relatively low mileage covered by buses and their robust design, but also perhaps economic factors which restrict the scope for capital investment in buses.

4.29   The comparisons above lead us to the following conclusions. Raising the standards of emissions control achieved by vehicles in service, in order to speed up the reduction of total emissions and to keep them low, is desirable in the case of all heavy duty vehicles. It is especially important in the case of buses. For many goods vehicles the relatively short operational life may prevent the benefits from outweighing factors such as cost and practicability. The importance of buses is reinforced by their place and mode of operation, as noted in Chapter 1. These points are reflected in the discussion which follows.

Retrofit

4.30 New components or equipment may be fitted to an existing engine or be added to the exhaust system. For the engine itself, this could for instance be in the form of improved fuel injector nozzles which reduce emissions of unburnt hydrocarbons. This may lead to a worthwhile reduction in emissions. If a package of such measures enabled an engine to meet the emission values specified in a stage of EC legislation beyond the stage whose values it met previously, the vehicle should qualify for the appropriate lower rate of vehicle excise duty which we recommend in paragraph 3.41.

4.31 For the exhaust system, aftertreatment may be added in the form of a flow-through catalyst or a particulate trap. The flow-through catalysts now being developed (paragraph 2.10) will be used to control emissions of hydrocarbons which are emitted in substantial quantities by some engines. They may be fitted to existing vehicles but most will require the use of low sulphur fuel to avoid the creation of unacceptable levels of particulates (paragraph 5.5). There is also a possibility of a catalyst which would control emissions of NOx; this would be of considerable benefit in reducing emissions from today's vehicles.

4.32 Most particulate traps (paragraph 2.9) are much more bulky and expensive than flow-through catalysts. This makes it difficult to argue for their widespread retrofitting, especially to most goods vehicles, but a strong case can be made for some classes of vehicle to be fitted with them. Buses appear to be especially appropriate. Trials have been conducted with buses in many countries, including very extensive trials in Germany[16] and in Greece[15,59]. In the German trial, 1100 buses and other vehicles had (at April 1991) been fitted with traps, mostly to existing vehicles. Up to 50% of the cost is met by the Government. The traps, almost all of which rely on catalysis for regeneration, are reported to have proved to be very reliable in service and to have created no operational difficulties.

4.33 We understand that, in this country, the Department of Transport is studying experience elsewhere and is planning to initiate trials[3]. A recent attempt to promote the use, in the UK, of a thermally regenerated trap manufactured abroad foundered because it was unable to operate with fuel containing the levels of sulphur at present supplied here[76]. This does not appear to pose a problem for all traps, however[13,16]. We recommend that the Government should proceed urgently with trials of particulate traps, concentrating on their application to buses. In the light of the extensive experience in other countries it should be possible for the UK to move rapidly to the implementation of an effective programme, including retrofitting. As in Germany, a substantial grant towards the cost should be offered for any vehicle taking part in the trials, especially because traps are at present expensive. Assuming that the trials prove the feasibility of retrofitting traps, a grant at an appropriate level should then to be offered for their retrofitting to any bus.

Rebuild

4.34 It is possible to rebuild a diesel engine, replacing some components and improving the condition of the whole engine, so as to return it to an 'as new' condition. An indication of what may be involved in this is in the box.

ENGINE REBUILD

In the rebuilding process offered by one major supplier([8]):

* The engine is stripped, cleaned, inspected, crack-detected, pressure tested and checked for dimensional accuracy.

* Cylinder blocks and cylinder heads have plugs removed, water passages cleaned and liners replaced and ground.

* Valves and crankshafts are checked, reground and polished.

* Fuel pumps are stripped and cleaned; fuel delivery is set to the same specification as for new engines.

* Fuel injectors are stripped, cleaned, reassembled using new parts where necessary and adjusted to ensure correct spray pattern and minimum fuel consumption.

* Turbochargers are stripped and cleaned, wheels and shafts reworked and fitted with new bearings.

* Water pumps are dismantled and cleaned, bearings and seals replaced and the units pressure tested.

* Pistons, rings, liners, bushes, bearings, gaskets, seals and many other parts are replaced.

* For the 'green rebuild' (paragraph 4.35), any necessary updating of the engineering specification of components is carried out.

* The rebuilt engine is tested in the same way as a new engine and carries a three year guarantee.

It is fairly common for the engines of buses to be rebuilt in this way at least once, in order to improve power output, fuel consumption and reliability. The rebuild may be carried out by the operator if they have the facilities, by an engineering company or by the original manufacturer. The effect on emissions is, broadly, to restore them to the levels to which the engine was designed.

4.35 One of Britain's largest manufacturers of bus engines is now offering a rebuild which goes somewhat beyond the ordinary one([8]). This so-called 'green rebuild', details of which are also in the box, is guaranteed to equip the engine to meet the emissions requirements of EC Directive 88/77/EEC. This imposes limits on emissions of carbon monoxide, unburnt hydrocarbons and nitrogen oxides (NOx), as specified in Table 3.1. For the first two, this guarantee is of limited significance, as an old engine even in poor condition would probably have been able to meet the Directive's requirements. For NOx, however, it represents a significant tightening of emissions control, down to little more than half of what might have been emitted by the engine previously and substantially better than would have been achieved by a standard rebuild. The cost is very little more than that of a standard rebuild.

4.36   The rebuild, even if 'green', is not guaranteed to reduce emissions of particulates, which are not covered by the present Directive. As noted in paragraph 4.2, however, particulate emissions tend to increase as the condition of the engine deteriorates with use. By restoring the engine to its 'as new' condition, therefore, the rebuild should lead to an improvement in particulate emissions control to its original value. The use of novel technology, such as improved fuel injection and piston design, could lead to significantly reduced particulate emissions.

4.37   The practice of rebuilding engines, and especially the 'green rebuild', is worthwhile. We recommend in paragraph 3.41 that the rate of vehicle excise duty (VED) should differ in accordance with the degree of emissions control achieved. This would be based on the certificated performance of the engine; it could thus apply as much to rebuilt engines in vehicles in service as to engines in new vehicles. EC Directive 88/77/EEC and its agreed amendment may not apply to rebuilt engines, in which case the application of the proposed varied rates of VED would not be subject to the constraints described in paragraph 3.42. A vehicle whose engine has been rebuilt to meet the emission limit values specified in the Directive, or in future the values in one of the stages of amendment, should qualify for the appropriate lower rate of VED. It would be necessary for the Government to be satisfied that the required standard of emissions control had been reached, so the work might need to be carried out only by the manufacturer or by accredited engineering works or operators; this would fit well with the pattern already established in the industry.

4.38   It is necessary to consider separately the application of this proposal to buses and to heavy goods vehicles. For buses, it is common practice for the engine to be rebuilt at least once. The additional cost of rebuilding to the requirements of the EC Directive 88/77/EEC, as compared with a standard rebuild, is small, so it seems likely that only a modest incentive need be created in order to persuade most operators to opt for it rather than for the other. The rate of VED for buses is low — only £450 per year for the largest — so the incentive which could be created by varying this rate is small. It should nevertheless be sufficient for this purpose. It seems less likely, however, to persuade many operators to bring forward the timing of a rebuild significantly, still less to have one done if they would not otherwise have considered it worthwhile. In view of the importance that we attach to the tight control of emissions from buses, we therefore recommend that an additional incentive be offered for them in the form of a grant for the fitting of a rebuilt engine which meets the emission values specified in the Directive, or in future the values in one of the stages of amendment, provided that the old engine is either scrapped or surrendered to be rebuilt to meet those values. It is necessary to add this proviso because a diesel engine, being very robust, might otherwise continue to be used for many years in another vehicle.

4.39   For heavy goods vehicles, engine rebuilds are rare, partly reflecting the generally shorter operational life of such vehicles. A larger incentive would therefore be required in order to persuade many operators to consider it and to stimulate a new market in offering rebuilds. On the other hand the VED for an HGV is much higher — about £3,000 for the heaviest — so a relatively large incentive could be created by varying this rate. This may be sufficient to persuade the operators of some of the longer lasting vehicles to have their engines rebuilt to meet the values specified in one of successive stages of EC legislation, which would meet our main objective.

Replacement

4.40 An engine may be replaced by one of improved design, as distinct from the rebuild described above, though this is much less common. It offers another route to improved emissions control, so that for instance a vehicle whose engine was not capable of being rebuilt to meet the emission values specified in the present EC Directive 88/77/EEC might be given a replacement engine which met those values. This would, of course, be more expensive than a rebuild and the old engine might have little or no surrender value.

4.41 Engine replacement is a practice which we consider to be worth encouraging. The new engine would attract the appropriate lower rate of vehicle excise duty recommended in paragraph 3.41. If it were possible, say, to replace an engine not meeting even the emission values of the present Directive with one meeting the values of Stage I of the agreed amendment, the resulting reduction of VED could be quite substantial for a goods vehicle. For a bus, because the VED reduction would be small (paragraph 3.47), a grant should also be offered, making this an attractive alternative to rebuild for some operators. This would be on condition that the old engine were either scrapped or surrendered to be rebuilt to meet the emission values specified in the Directive, or in future the values in one of the stages of amendment, for use in another vehicle. The size of the grant should reflect the certificated emissions performance of the replacement engine: one amount if the engine met whatever were the limits then in force and a larger one if it met limits which had been adopted but not yet implemented.

CHAPTER 5

FUELS AND LUBRICANTS

Introduction

5.1   Diesel fuel is a blend of hydrocarbon streams from the distillation of crude oil and other refinery processes (see box on page 50), differing widely in chemical composition and hence in physical properties. There are smaller amounts of other substances, notably sulphur and traces of metals. This Chapter considers the impact upon emissions of changing or maintaining a number of characteristics of fuel, the costs involved and other implications. The effects of additives for fuel and lubricants are also considered. Finally there is a brief assessment of the prospects for replacing diesel with petrol or other fuels.

Fuel Characteristics and Emissions

5.2   Amongst the characteristics of diesel fuel which are commonly specified are:

— cetane number, a measure of the ease of ignition of fuel in an engine;

— chemical composition such as sulphur content; and

— physical properties such as boiling point and density.

5.3   The relationship between diesel fuel characteristics and emissions has been investigated in a large number of studies. These have been largely empirical, with only limited understanding of the combustion processes which give rise to the emissions. Changes in fuel characteristics have been shown to have some impact on emissions of NOx, hydrocarbons, particulates and smoke. The scale of the effects has varied but a consensus of views of the engine manufacturers and the oil industry has been reached within the forum of the Motor Vehicle Emissions Group of the European Commission. After a comprehensive study of the literature[78] the following trends were identified:-

> There was a modest increase in particulates and hydrocarbons with decreasing cetane number and increasing boiling point. Increases in particulates ranged from 5 to 20% as the result of a decrease of six in cetane number, or an increase of 20°C in the 85% boiling point (see box on page 50).

> Particulates were decreased by a decrease in sulphur content.

> No consistent effect was found in NOx emissions as the result of changes in cetane number or boiling point.

> No effect on emissions of a change in aromatics content was found, beyond that linked with the associated change in cetane number.

With reference to the effect of cetane number and volatility on particulates, it should be noted that the lower end of the cetane range used in the tests, 44, represents the lower end of the range of cetane numbers of the fuel supplied in Europe. The bulk of supplies, however, fall within the range 48-50. The 85% boiling point also normally varies within a much smaller range than 20°C. The variations in emissions resulting from the bulk of fuel supplied are therefore much less than the 5–20% recorded in these tests.

THE PRODUCTION OF DIESEL FUEL

Diesel fuel is refined from crude petroleum oil. Crudes from different parts of the world differ considerably in the types and proportions of hydrocarbons they contain. The molecules of each hydrocarbon have the atoms of hydrogen and carbon arranged in a characteristic pattern. The main types in crude oil are paraffins, containing straight or branched chains of carbon atoms, and aromatics, containing one or more benzene rings of carbon atoms in addition to chains. Paraffins tend to have better properties for use in diesel fuel than do aromatics. The molecules also differ in the proportions of the two elements that they contain: the lightest, with high proportions of hydrogen atoms, are methane, propane and butane, whilst the heaviest residue oils contain high proportions of carbon atoms.

The primary process used in an oil refinery is fractional distillation. The components of crude oil have different boiling points; in general the smaller the molecule the lower the density and the boiling point. The main components, or 'fractions', in ascending order of boiling point, are: petroleum gases (propane and butane); light and middle distillates used as components of petrol, petrochemical feedstocks, aviation turbine fuel and gas oil (diesel fuel and heating oil); and heavy distillates which are processed further to produce lubricants and waxes. Diesel fuel has an 85% boiling point (the temperature at which 85% of it is vaporised) of about 350°C.

Crude oil is distilled by heating and partially vaporising it in a tower containing equipment which brings the ascending vapours into contact with descending liquid. The lightest fraction leaves the top of the tower as vapour, to be partially condensed or taken off as a gas. Part of the condensate is returned to the tower, where it descends, initiating the condensation of successively heavier fractions at successively lower levels of the tower; these fractions are withdrawn for blending. into products or for further processing. The non-volatile residue of the crude oil is withdrawn from the bottom of the tower and may be distilled under vacuum.

Fractional distillation is a physical process which does not alter the composition of the hydrocarbon molecules. Other refining processes are used to bring about chemical changes. Heavy distillates and residues of the vacuum distillation process are 'cracked' to shorten the chains of carbon atoms and thus to create lighter products. This enables the range and volume of products manufactured to be varied to match the market demand for them. The first cracking processes used high temperature and pressure alone to effect this chemical transformation and are called thermal cracking or coking. They have largely been replaced by methods in which the molecular breakdown is promoted by catalysts; these are known as catalytic cracking. Some modern refineries, particularly in the USA, carry out the conversion process using hydrogen under pressure and in the presence of a catalyst; this is known as hydrocracking or hydroconversion. Its products tend to be low in aromatics, so that the appropriate ones are high quality components for diesel fuel. Hydrocracking is not yet widely used in Europe.

The concentrations of sulphur and aromatics in the various fractions can be reduced by a form of catalytic conversion process, in the presence of hydrogen, under varying conditions of temperature and pressure.

Diesel fuel is usually a blend of fractions derived from crude oil distillation and certain products of secondary conversion processes, as indicated in Figure 5.1. Fuel supplied in Europe is composed about 80% of the straight run fraction from crude distillation, the remainder being from catalytic cracking with small amounts from thermal or hydro-cracking. These have differing qualities as components of fuel, which can result in variations in the characteristics of diesel fuel supplied from different refineries.

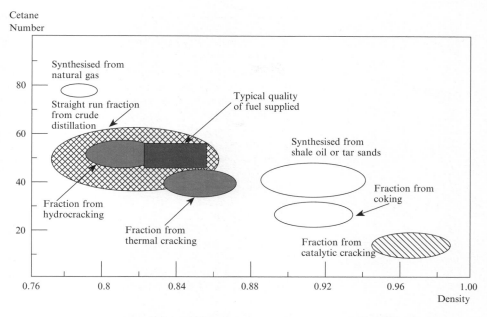

Figure 5.1: Typical Properties of Middle Distillate Fractions Used in Diesel Fuel([77])

5.4   The potential for reducing emissions by changing fuel characteristics has been shown to be small in comparison with the reductions effected by improvements in engine design in recent years. It may nevertheless have a part to play as further advances in engine design become increasingly difficult and costly to attain. Regulatory authorities and the industry are giving increasing attention to the control of fuel characteristics such as sulphur content, cetane number and aromatic content. These are considered in turn below.

Sulphur

5.5   The strongest and most consistent relationship between an aspect of fuel and emissions is that between sulphur content and particulates. The conversion of fuel sulphur into particulate emissions is described in paragraph 2.4. Typically about 2% is converted, the remainder being emitted as sulphur dioxide. The use of most oxidation catalysts to control emissions of hydrocarbons has the incidental effect of sharply increasing this conversion rate and hence of increasing the mass of particulates recorded in the engine test, though development work on catalysts is reducing the severity of this. In many catalytic traps the additional particulate material can lead to the trap's becoming clogged.

5.6   The maximum permitted concentration of sulphur in diesel fuel in the European Community is at present 0.3%. It has been agreed that this should be reduced to 0.05% by 1996 as an adjunct to the limit on particulate emissions agreed for that date. The concentration depends upon the composition of the crude oil and the secondary processing used in the refinery. The UK has been fortunate so far in having access to relatively low sulphur crude oil from the North Sea, Nigeria and Algeria. As supplies from these sources decline they are likely to be replaced by heavier, high sulphur crudes from the Middle East and the western hemisphere. Sulphur can be removed by the reaction of oil with hydrogen in the presence of a catalyst under varying conditions of temperature and pressure. The lower the sulphur content required the more extreme the operating conditions need to be. A reduction to 0.05% will require the construction of additional plant in European refineries.

5.7 The cost of reducing the sulphur content of diesel fuel to 0.05% in Europe was estimated in 1989 by the consultants A D Little, on behalf of the German Federal Environmental Agency, and by CONCAWE on behalf of the oil industry[78]. The estimates differed widely, mainly because of disagreement over how much additional desulphurisation capacity would be needed. They concluded that additional capital expenditure of $470 million (Little) or $2700 million (CONCAWE) would be required, with an increase in the pre-tax price at the pump of 1p or 4-5p per gallon. We understand, however, that discussion has led to a narrowing of this difference towards the lower part of the range[16]. There will also be an environmental cost in terms of consumption of energy in the refinery and hence the emission of carbon dioxide. An additional 0.8-1 million tonnes of oil will be required to process the 78 million tonnes of diesel oil expected to be consumed annually in Europe by 1995[79].

5.8 The reduction in the maximum permitted fuel sulphur content is to be effected by an EC Directive. The European Commission's proposal[80] would require diesel fuel to have no more than 0.2% sulphur from October 1994 and 0.05% from October 1996, when the Stage II limits on engine emissions (paragraph 3.7) come into effect. In addition, Member States would be required to ensure the availability on the market and a balanced distribution of 0.05% sulphur fuel by October 1995. To this end, Member States are encouraged to introduce a tax incentive for 0.05% sulphur fuel in 1992. We support these proposals. We recommend that the UK Government should introduce a fuel duty differential in favour of low sulphur fuel which is more than sufficient to off-set the additional cost of its production, thus compensating the producer and enabling the retail price to be lower. It should do so from the earliest possible date and should maintain the differential until the use of such fuel becomes mandatory.

5.9 We wish to encourage bus operators to start using the low sulphur fuel as soon as it is available. They receive a rebate of fuel duty from the Government, so they would not be influenced by a change in the duty. We recommend, therefore, that the Government should consider restricting this rebate to diesel fuel with a sulphur content not exceeding 0.05% at the earliest practicable date. This would create a powerful incentive for its use. Buses obtain their fuel from depots, so it should be possible for operators to obtain the new fuel before it is distributed widely to service stations.

5.10 The proposed amendment to the Directive also provides for maximum permitted sulphur concentrations for diesel oil supplied for off-road uses and for heating. That also would be set at 0.2% from October 1994, followed by 0.1% from October 1999. In view of the contribution which this should make to the control of emissions from off-road uses, referred to in Chapter 3, we support these proposals also.

Cetane Number

5.11 Most diesel fuel supplied in Europe has a cetane number in the range 48–50. The British Standard for diesel fuel[81], which is not mandatory, had until 1989 specified fuel with a minimum cetane number of 50. After strong debate between fuel suppliers and vehicle operators this was reduced to 48, to allow for the effects of forecast changes (described below), though we understand that market pressures have so far resisted any change in the cetane number of fuel supplied. The European Committee for Standardisation (CEN) has recently drafted a proposal to establish a European Standard for diesel fuel which would include a minimum cetane number of 49[82]. This would require the British Standard to be changed again, to come into line with it, but it would still not be mandatory. This contrasts with the position on standards for petrol, which in this country have for many years been the subject of legal controls and are now covered by a European Directive.

5.12 Diesel fuel supplied to the UK market is blended from the products of the distillation of crude oil and of secondary conversion processes commonly available in north west Europe, as described in the box on page 50. The decline in demand for fuel oil, relative to the other refined fuels, has led to increased use of the products of secondary conversion processes and diesel fuel receives increasing amounts of them. With the exception of the products of hydrocracking (see box), these generally have a lower cetane number and a higher aromatics content than the fractions of crude oil distillation. The cetane number of diesel fuel from the refinery is expected to fall during the next few years if the trends in markets and the current selection of refinery processes continue ([79,83]).

5.13 The operation of refineries is flexible and can, within certain limits, be adjusted to compensate for changes in fuel components. Better segregation of gas oil distillates, for instance, may be used to route the higher cetane components for use in diesel fuel and the lower ones for heating oil. Hydro-desulphurisation (paragraph 5.6) of diesel fuel also has the effect of raising cetane number slightly. The addition of desulphurisation to the refining process for diesel fuel in Europe over the next few years should reduce the effect of the trends referred to above but is not likely to eliminate it. If any additional cracking capacity required in Europe over the next decade were to be provided in the form of hydrocracking, rather than catalytic and thermal cracking processes which have been used hitherto in Europe (see box on page 50), the result may be to slow or even halt the decline in cetane number from the refinery.

5.14 Additives may be used to raise or maintain cetane number, offering a very much cheaper and more flexible means of doing so than processing oil with hydrogen([79]). The increasing use of such additives has partially offset a fall in cetane number of fuel from the refinery in the past few years([83]). There is a possible disadvantage to such an approach however. A fall in cetane number is frequently associated with a rise in aromatics content so that the increased use of additives to maintain cetane number could mask an increase in aromatics in the fuel. The possible effect of this is described below. To guard against this, it has been proposed by CEN that the European Standard referred to in paragraph 5.11 should, in addition to the cetane number, specify a minimum cetane index (see glossary), a narrow density range and a minimum 95% boiling point. This would have the effect of preventing any significant increase in aromatics content which, in turn, would limit the extent to which additives would be needed to maintain cetane number.

5.15 The variation in emissions with cetane number, reported in paragraph 5.3, arises from differences in cetane number between the fuel supplied in the tests and that which the engine was designed to use. The 'homologation' (certification) fuel for the European Community's engine emissions test, described in paragraph 3.22 and the box, has cetane number 50. Vehicles in service, as well as new ones, have been designed to run on such fuel. A drop in the cetane number of fuel supplied would therefore increase the emissions from all vehicles. The lack of legal control over diesel fuel characteristics is anomalous and has no obvious justification. We recommend that the UK Government should consider introducing legal control of the cetane number and perhaps other characteristics of diesel fuel. It should also consider proposing that an EC Directive on this be introduced.

Aromatics

5.16 Hydrocarbons containing aromatic rings typically constitute 20–35% by volume of diesel fuel supplied in Europe. It is possible that there is a relationship between their level in the fuel and the level of polyaromatic hydrocarbons (PAHs) and particulates in the exhaust emissions. It is not clear, however, how different aromatic compounds behave in the combustion process[83]. Further understanding of this is important because some PAHs are carcinogenic (paragraph 2.27). There are several difficulties in the way of determining the effects of aromatics in fuel:-

(a) There is no agreement on which compounds in diesel fuel should be classified as aromatic or on tests for aromatic content. Movement is being made in this direction, however, and the European Committee for Standardisation is evaluating proposals for test methods.

(b) Some fuel characteristics are interrelated and it is difficult in practice to set up test fuels for comparison which differ systematically and separately in aromatic content, cetane number, boiling point and density.

(c) The techniques used to analyse emissions in the certification test measure total unburnt hydrocarbons. The same techniques are commonly used in experimental studies of engine performance. Few studies have been made of the amounts of specific compounds in the emitted hydrocarbons.

The aromatics content of diesel fuel may be reduced by the use of selective hydroprocessing but this could involve much higher costs than will the reduction in sulphur content to 0.05%[82]. A reduction might be achieved by using only selected distillation products but this could result in a severe loss of flexibility in the operation of refineries if applied to the bulk of production [84].

5.17 Some authorities are already seeking to control the aromatics content of diesel fuel. The California Air Resources Board has proposed a limit of 10% by volume on the aromatics content of engine homologation fuel from October 1993[85], though it is relevant that the composition of diesel fuel in the USA makes a test of the aromatics content more reliable than for fuel supplied in Europe and the removal of aromatics less costly[82]. We have received evidence from the representatives of European Community vehicle manufacturers[57] and UK operators[14,86] advocating that the aromatics content of diesel fuel in Europe should be subject to a similar limit. Some other EC Member States are considering the merits of such an approach[16].

5.18 The presence of aromatics in diesel fuel is a source of concern and there is a prima facie case for their limitation in order to reduce the formation of PAHs in exhaust emissions. We consider, however, that it is necessary first to determine with greater certainty, and in more detail, the relationship between defined aromatic compounds in the fuel and the composition of emissions and to assess the environmental hazards which the emissions present. The UK Government and the European Commission should commission such research. If it is shown that specific aromatic components in diesel fuel result in harmful emissions, the concentration of those components in the fuel should be limited. This would stimulate the development of refinery techniques which would discriminate between different types of aromatic components and would reduce only those which are the precursors of harmful emissions.

Additives

5.19 Additives are being increasingly used both to improve the performance of diesel fuel as currently formulated and to compensate for changes in fuel, especially as cetane improvers (paragraph 5.14). Lubricants also contain high proportions of additives. Atomisation and combustion of the fuel and the reduction of deposits can all be improved by the use of detergents and cetane improvers as additives. Experiments with test engines are claimed to show significant decreases in emissions, especially unburnt hydrocarbons (20–40%) and particulates (10–13%)[87]. If these effects were reproduced in working engines they would be substantially greater than the improvements in emissions reported as a result of other changes in fuel characteristics apart from a reduction in sulphur content. The decrease in NOx emissions, 5% in the experiment quoted, is much less; we were told that the development of an additive which would effectively control NOx formation was only a remote possibility.

5.20 It has been put to us by industry representatives[87] that most additives used today are compounds of only carbon, hydrogen, oxygen and nitrogen and that the emissions resulting from their use are the combustion products of those compounds. We cannot, however, rule out the possibility that the mode of action of some classes of additive entails the release of products which are a risk to health. For example, the nitro-groups used in cetane improvers such as alkyl nitrate might contribute to nitrogenated PAHs which are carcinogens. It is important to seek to ensure that the use of additives to control the emissions from diesel engines does not introduce a secondary emission problem from the combustion of the additives themselves including 'cocktail' effects in the exhaust. Toxicological studies of additives that are currently undertaken relate only to the handling of the bulk material by industry and its ecotoxicity. As far as we are aware little or no toxicological testing has been carried out on the combustion products of additives, although work has recently started in the UK[87], the USA[64] and Germany[16].

5.21 We were told that, in general, metals are not used as diesel fuel additives in the UK[87]. Metals are, however, included in some additives which the operator may add to the fuel in a vehicle's tank; these are probably manufactured only in small quantities. They are also used in some lubricants. In addition, metallic or organo-metallic additives may soon be used in a variety of particulate traps. They would be added to the fuel or the exhaust stream to induce regeneration of the trap by the burning of the particulates. One employing copper (see box with paragraph 2.9) is on trial in the UK and one using cerium has been developed on buses in Greece and is now going into production [15, 59]. In Germany, some additives contain metals[16].

5.22 Powers are available under the Environmental Protection Act 1990 for the control of the use of injurious substances and the provision of information. Regulations are required for each specific substance and the Secretary of State would require good reason to believe that the substance presented an environmental hazard before acting[30].

5.23 The use of diesel fuel additives is growing and it can be expected that even greater use will be made of them in the future. We view with concern the lack of knowledge of the emissions to which these substances may give rise. The Government should use its new powers to control the use of fuel additives to the full. We recommend that it should adopt the following approach:-

The use of potentially hazardous metals as fuel additives should be banned until the combustion products of such additives, emitted from the exhaust, have been subjected to appropriate toxicological testing.

No new substance should be permitted to be used as a fuel additive until similar testing has been carried out on it.

A programme of such testing on existing substances used as fuel additives should be established as a basis for reviewing their continued use.

If a metal additive is used for regeneration of a particulate trap and most of the metal is retained within the filter element, this should be taken into account in evaluating the environmental impact of the emissions. If the metal is potentially hazardous, care should be taken in the eventual disposal of the filter.

Further Improvement in Diesel Fuel

5.24   The reduction in the sulphur content of diesel fuel to 0.05% will represent an important improvement. The use of such fuel will be required by 1996. There may then be a case for introducing a further improvement in diesel fuel. This would probably include a limit on the total aromatics content, or on certain aromatic components, if research showed it to be justified. A higher cetane number would probably result from any reduction in aromatics content though there is at present little evidence to indicate that this would be of additional environmental benefit. The merits of introducing such an improved fuel should be kept under review, taking account of the energy cost of its production.

5.25   It seems likely that all diesel vehicles would be able to run on such an improved fuel. If one is introduced, therefore, its use should be required from an appropriate date. That date might be set several years ahead, to allow time for the necessary investment in additional refining plant. In the interim, financial incentives should be created for the use of the new fuel similar to those we recommend for 0.05% sulphur fuel (paragraphs 5.8 and 5.9).

5.26   The basis on which a standard for diesel fuel is specified also merits attention. The present specification relates to the fuel's physical properties and its sulphur content. It has been suggested to us[88] that a better approach would be to define the standard of a fuel by its emissions performance in a reference engine under controlled conditions. Fuels having different chemical composition and physical properties might be permitted, provided that the specified emissions performance were achieved. This would allow each fuel supplier to use the most cost-effective way of reaching the required emissions performance, subject to the customer's requirements that the fuel should meet other performance characteristics such as cold-weather starting.

5.27   There are reasons to be cautious of such an approach. First, it depends upon the supposition that the emissions performance of most engines respond to changes in fuel characteristics in the same way, so that a 'typical' engine may be used as a reference. Our technical advice is that this cannot by any means be taken for granted[66]. Second, the adoption of such an approach may face problems of compatibility with production control in the refinery. Also, variations in the chemical composition and physical properties of the fuel may give rise to emissions problems, as exemplified by the possibility of an increase in aromatics content being masked by increased use of cetane number improvers (paragraph 5.14). Much greater understanding of the combustion processes involved in a diesel engine, and more discriminating techniques for measuring emissions with correspondingly selective limit values, would be needed before such an approach could be unequivocally endorsed. We have recommended in Chapters 2 and 3 that such techniques should be developed and such limit values adopted. When this has been done, the Government should consider whether the development of a diesel fuel standard related to emissions performance is then feasible and desirable.

Lubricants

5.28 Lubricating oil passing the piston rings and valves into the combustion chamber is a significant source of the hydrocarbon component of particulate emissions, contributing as much as 25% of the total in engines of current design. One way to control these emissions is to reduce the amount of lubricant consumed. In doing so it is necessary still to supply sufficient for adequate lubrication of the rings and valves. Design improvements are underway to achieve this[89].

5.29 Another approach to the control of emissions arising from lubricants is to change their formulation. Development of low emissions lubricants is an area of active research and it is expected that a number of products for heavy duty diesel vehicles will be marketed within the next few years[88]. When available, such lubricants are likely to be used by engine manufacturers in their certification tests. There is at present no legislative control over the composition of lubricants or their emissions performance. Such control should be considered as an adjunct to the standards for particulate emissions agreed for the European Community from 1996. Because lubricants are complex mixtures of blended oils and additives, and are subject to tight commercial secrecy, a standard based on composition might not be appropriate. Investigation of an alternative basis for a standard, related to emissions performance, should be undertaken.

Alternative Fuels

5.30 Lower emissions of particulates, NOx and in some cases hydrocarbons, could be achieved by the use of spark ignition engines with three-way catalysts, burning petrol or another fuel. In Britain and the rest of Europe, petrol engines powered the heavy duty fleet until displaced by diesel engines during the late 1940s and the 1950s. The principal reason for the petrol engine's decline in Europe is that its fuel consumption was (and still is) higher than that of a diesel engine of equivalent power. In the USA a high proportion of medium heavy duty vehicles have petrol engines, due partly to a low fuel tax. The petrol engine's lower fuel efficiency, combined with the substantial additional refining capacity and energy consumption which would be needed to convert into petrol the oil fraction which at present is blended into diesel fuel, suggests that the petrol engine will not regain use in heavy duty vehicles unless strongly encouraged by governments. We do not consider that such encouragement would be worthwhile, except for a limited number of applications as indicated in paragraph 5.34.

5.31 Several other fuels are being considered for use in spark ignition engines, either in new vehicles or as replacements for diesel engines. They include methanol, compressed natural gas (CNG), liquified petroleum gas (LPG), hydrogen and a range of biomass-derived fuels such as ethanol and rape oil. Most of these present problems of safety, cost or large scale fuel supply which make it unlikely that they will be contributing significantly to the heavy duty vehicle market in the foreseeable future. In the cases of CNG and LPG, however, there is an existing core of users and adequate supplies are available. Vehicles running on CNG or LPG have been in use for a number of years in many European and other countries, partly because of a fuel duty regime which favours them[90]. Proposed harmonisation of duty rates in the European Community could make these fuels even more financially attractive to vehicle operators in the UK. An initial pilot project is underway in Blackburn where a number of Council vehicles are running on CNG[91].

5.32   The principal advantage of using CNG or LPG rather than diesel is the reduction in certain emissions. It has been reported[90] that, as compared with a diesel engine of equivalent power, a CNG or LPG engine with a catalyst emits only about 7% the mass of particulates, 4% the hydrocarbons and 35–40% the NOx. There are drawbacks, however:-

> Both fuels have a low density, restricting the operating range of the vehicle or requiring more frequent refuelling. Other operating difficulties stem from the need to store and carry the fuel under pressure.

> The cost of replacing engines and fuel tanks is high.

> CNG consists largely of methane, a powerful 'greenhouse' gas. Leakage of the fuel, from the vehicle or from the distribution network, will release it into the atmosphere.

It appears unlikely that these fuels will gain widespread use in the near future. We consider their use to be desirable in a limited number of applications, as indicated in paragraph 5.34.

5.33   The use of electric batteries for vehicle power offers the possibility of virtually zero emissions at the point of use. Such vehicles have been in use in this country in certain specialised applications, notably as milk floats, for many years. Advances in battery design in recent years, for instance the development of the sodium-sulphur cell, have led to a reduction in their weight and an increase in the potential operating range and speed of the vehicle[92], though these are at present still inferior to those of most other sources of vehicle power. The supply of electric power through overhead cables or through guide rails has also been used for buses and trams in urban areas but these lost favour. Electric battery vehicles for use in towns are becoming increasingly practicable and pilot production of vans and small buses was undertaken in Britain in during the late 1980s. The generation of the electrical energy adds to the emissions from power stations and is subject to the usual transmission losses. Emissions are generated outside the areas of high vehicle concentration, however, and derive from point sources which may be effectively controlled. It appears that this is an area of considerable promise for some applications but that its widespread application to heavy duty vehicles during the next decade is highly unlikely.

5.34   We consider that the most favourable sector for the use of alternative fuels over the next decade is the bus. In the case of LPG the lack of a retail distribution network makes it appropriate for use only by vehicles operating from depots. CNG may be obtained from the mains gas supply, though compression equipment would be required. We recommend that the Government should conduct a study of the effect of the production and use of alternative fuels on carbon dioxide and other emissions. Subject to its judgment of the conclusions of that study, it should seek to encourage the use of alternative fuels by buses and should consider the most appropriate means of doing so. Since bus operators have fuel duty rebated they will not be influenced by differential rates of duty. An alternative approach might be to offer a grant towards the cost of fuel storage and delivery equipment. We recommend that, provided that the vehicle met the emission values specified in one of the stages of EC legislation pertaining to diesel vehicles, the appropriate rate of vehicle excise duty recommended in paragraph 3.41 should be charged. It should also attract the appropriate level of grant for a new bus or a replacement engine recommended in paragraphs 3.48 and 4.41 respectively.

CHAPTER 6

THEMATIC SUMMARY

6.1   This Chapter is a summary of the main points of the Report, grouped thematically rather than in chapter order. It identifies many of the key topics, pulling together points which have been made in one or more of the previous chapters. Recommendations are indicated by references in ( ) to their numbers in Chapter 7.

Emissions and Their Impact

6.2   The principal emissions from heavy duty diesel vehicles which are considered in this Report are particulates, nitrogen oxides (NOx), unburnt hydrocarbons and visible smoke. The greatest attention is paid to the first two. Particulates are composed mainly of carbon, hydrocarbons, sulphates and water. They result from the incomplete combustion of the fuel which is itself composed chiefly of hydrocarbons. Particulates are of concern because of the risk to human health, especially from the polyaromatic hydrocarbons (PAHs); they also soil and damage buildings. NOx has been shown to have an adverse effect on health and contributes to acid deposition; it is also a precursor of ozone which, in the lower atmosphere, is an irritant to breathing and damages trees and crops.

6.3   Most emissions from heavy duty diesel vehicles have been rising in recent years, reflecting an increasing number of vehicles and distance travelled by them and the use of more powerful engines. Road traffic as a whole contributes nearly half the national total emissions of NOx and diesels account for nearly half of that. Diesel vehicles are now the major source of smoke in urban areas. As the use of diesel vehicles continues to grow, the need to achieve effective control of their emissions becomes increasingly important.

The Control of Emissions from New Vehicles and Engines

6.4   Modern diesel engines have lower emissions per unit of power than hitherto. The number of vehicles emitting visible smoke has declined sharply. This has been achieved by a variety of developments in engine design and component specification and further developments are to be required within the European Community (EC). An EC Directive on emissions from heavy duty vehicles came into force in October 1990 and an amendment to it has been agreed by the EC Council of Ministers. The existing Directive imposes limit values on emissions of NOx, hydrocarbons and carbon monoxide. The amendment tightens these limits in two stages, taking effect in 1993 and 1996, and adds a limit on particulate matter. We support the view of the UK Government that the costs of compliance with the emission values specified in this legislation are justified by the benefits to the environment.

6.5   The degrees of reduction which can be achieved for different emissions are linked. Most importantly there is a trade-off, for any one engine type, between emissions of NOx and of particulates. Until now the emission limits required by EC legislation appear to have been based largely on technological considerations, taking no explicit account of this trade-off and of the relative priority which should be accorded to the control of different emissions. A more fundamental approach should be taken to setting emission limits for vehicles. Guide values for air quality should be set in the light of the best available data on the health and other impacts of pollutants. Measures should be devised for moving towards achievement of the guide values, having regard to the growth in vehicle numbers and use and to other sources of the relevant pollutants. The Government should commence the necessary programme of work (7.4).

6.6   Until that work has been done, however, the present approach to setting emission values is the most appropriate course of action in the light of present knowledge of the environmental impacts of the various pollutants (7.5). Future emission limits for heavy duty vehicles in the EC, including those to be implemented at the end of the decade, should require further reductions in emissions both of NOx and of particulate matter. This policy will need to be reviewed (7.6). Consideration should be given to the setting of separate limit values for some of the various components of diesel emissions (7.7).

Engine Test Cycles

6.7   The UK Government has proposed that the test cycle used to measure the emissions from engines for the purposes of certification in the EC should be changed. In particular, it has advocated adoption of the transient cycle used in the USA, in which the engine is taken through a controlled pathway of changes in speed and load, its emissions being measured continuously. This would replace the steady state test used at present in the EC and elsewhere, in which the engine is run at a series of different steady state modes, its emissions being measured only when at those modes.

6.8   We consider that a transient test is, in principle, likely to provide a better indication of emissions performance on the road, especially for modern diesel engines. The Government was right to propose adoption of the US test for the amendment to EC Directive 88/77/EEC. In the event, it has been agreed that no change shall be made for the present, but the issue is to be reviewed. There would be value in achieving wide international harmonisation in the equipment used to test vehicle engines. The cycle of speed and load conditions over which the engine is taken, however, could be allowed to differ to represent the driving conditions of particular concern in each country or region. The UK Government should press the EC to develop an engine test which uses equipment to the same specification as the US transient test but whose cycle emphasises the representation of European urban driving conditions (7.9).

Economic Instruments

6.9   The control of emissions from vehicles, including heavy duty diesel vehicles, is well suited to the application of economic instruments. Pollution results from the cumulative impact of large numbers of vehicles, so there would be advantage in influencing a substantial proportion of operators to use less polluting vehicles even though a minority persisted in using more heavily polluting ones. Economic instruments could be used to speed up the introduction of the cleaner vehicles by encouraging manufacturers to produce engines which are able to meet new emission values before the date on which they are required to do so and by encouraging operators to purchase the new engines or vehicles.

6.10   A practical form of incentive for the manufacture and the purchase of vehicles with less polluting engines would be differences in the rate of vehicle excise duty (VED) according to whether the engine meets the emission values specified in successive stages of EC legislation. These rate categories could best be applied on the basis of the certificated performance of the engine. They should be applied for the whole life of the vehicle, unless and until it is fitted with an engine qualifying it for a different rate (7.12). The agreed amendment to the EC Directive 88/77/EEC allows for Member States to implement part of such a scheme. The UK Government should press for the Directive to be further amended to widen the scope for the application of economic instruments (7.13).

6.11 The emissions control achieved by a vehicle in service may be upgraded by a variety of means. One is to retrofit equipment, to the engine or for exhaust aftertreatment. Another is to rebuild the engine to its 'as new' condition or to an even higher standard. A third is to replace the engine with one of more modern design. If, as a result of any of these, the vehicle's emissions fall to the level specified in one of successive stages of EC legislation for new vehicles and engines, it should qualify for the appropriate lower level of VED.

6.12 The multi-rate VED would thus provide an incentive for the replacement or upgrading of vehicles and engines in service to achieve improved levels of emissions control and for the early production and purchase of engines achieving emission values adopted by the EC but not yet in force.

6.13 A grant should be paid towards the cost of a new engine in a new bus, meeting the emission values specified in one of successive stages of EC legislation, purchased before the relevant stage was implemented and the values became mandatory (7.15). A grant should also be paid when a bus receives a rebuilt or replacement engine which meets tighter emission values, provided that the old engine is either scrapped or surrendered to be rebuilt to meet the values specified in EC Directive 88/77/EEC or, in future, the values in one of the stages of amendment (7.19, 7.20). A grant should be offered for the fitting of a particulate trap to a bus, whether new or in service (7.14, 7.17).

6.14 The Government should introduce a fuel duty differential in favour of low sulphur fuel which is more than sufficient to off-set the additional cost of its production, until the use of such fuel becomes mandatory (7.21). It should consider restricting the fuel duty rebate for buses to low sulphur fuel (7.22).

The Urban Environment and Buses

6.15 Our objective is to ensure protection of the environment in all parts of the country and most of our recommendations are intended to result in improvements wherever diesel emissions may be a problem. The larger urban areas are subject to very high concentrations of vehicles and their emissions. In addition, the number of people and buildings exposed to the pollutants is greater in urban areas than elsewhere. We have therefore considered whether additional steps to control emissions in urban areas might be warranted. We have particularly considered buses, which operate mainly in urban areas and which may be kept in operation for many years. Consideration should be given to the setting of much lower limits for the emission of particulates and NOx from buses than from other vehicles at the end of the decade (7.24, 7.25). Other measures to control emissions from buses are described in the sections on economic instruments and vehicles in service.

6.16 Many of the considerations which apply to buses apply also, to varying extents, to other classes of vehicle which operate mainly in urban areas.

The Government should carry out the necessary work to identify the classes of vehicle which make the largest contributions to emissions in urban areas and should consider how our recommendations for buses might be applied to them (7.26).

6.17 Measures to reduce the use of vehicles in urban areas could also contribute to a decrease in emissions there. Such measures should be considered but they raise broader issues and do not fall within the scope of this Report.

Standards and Tests for Vehicles in Service

6.18 Requirements about emissions from a heavy duty diesel vehicle in service should be imposed relating to the same range of emissions as was specified in legislation when the vehicle or engine was new. It is impractical to measure the emissions from vehicles in service in a reliable way. It is therefore necessary to seek other ways of stating the requirement for vehicles in service to perform satisfactorily in respect of emissions control. One is to require that engines be designed not to deteriorate in emissions performance by more than a specified factor over prolonged periods of operation; such an approach should be considered by the UK Government for incorporation into the European engine certification procedures (7.27). It will shortly be possible to require that engines be kept in good condition and to apply diagnostic techniques to test this. The Government should develop this approach as part of its own legislative control and testing of emissions from heavy duty diesel vehicles in service and should propose its introduction throughout the European Community (7.29, 7.33, 7.35). We welcome the Government's intention to introduce an instrumented smoke check for heavy duty vehicles and to require the fitting of speed limiters to heavy goods vehicles.

6.19 The Vehicle Inspectorate carries out spot checks of emissions of smoke. More of these should be carried out and they should incorporate the new instrumented measure of smoke and diagnostic techniques to indicate performance in controlling other emissions (7.36). Some local authorities are introducing schemes to spot smoky vehicles. These are worthwhile and should be developed, with or without public involvement (7.39). The number of vehicles emitting visible smoke seems likely to continue to decline. Permanent arrangements should, however, be made to ensure that the public is aware of how to report smoky vehicles to the Vehicle Inspectorate (7.40). The offence of emitting smoke in such a manner that other road users are endangered remains a useful sanction against the worst cases of smoke pollution.

6.20 The Government should proceed urgently with trials of particulate traps, concentrating on their application to buses. A substantial grant towards the cost should be offered for any vehicle taking part in the trials (7.41).

Fuels and Lubricants

6.21 The characteristic of diesel fuel which has been demonstrated to have the most significant effect on emissions is the sulphur content. The maximum permitted content is to be reduced to 0.05%, as an adjunct to the new emission limits to be implemented in 1996. We support this. The merits of introducing a further improved diesel fuel should be kept under review (7.47).

6.22 It is important that diesel fuel should have a cetane number not significantly lower than that for which engines have been designed and certificated. The UK Government should consider introducing legal control of the cetane number and perhaps other characteristics of diesel fuel (7.42). The UK Government and the European Commission should commission research into the relationship between defined aromatic compounds in diesel fuel and the composition of emissions and to assess the environmental hazards which the emissions present (7.43). The use of potentially hazardous metals as fuel additives should be banned until the combustion products of such additives, emitted from the exhaust, have been subjected to appropriate toxicological testing. No new substance should be permitted to be used as a fuel additive until similar testing has been carried out on it and the continued use of existing substances should be reviewed on the same basis (7.45).

6.23   Petrol, liquified petroleum gas, compressed natural gas and electricity are possible alternatives to diesel fuel for heavy duty vehicles but none appears likely to gain widespread application over the next decade. The Government should conduct a study of their effect on carbon dioxide and other emissions (7.49). Subject to its judgment of the conclusions of that study, it should seek to encourage the use of alternative fuels by buses and should consider the most appropriate means of doing so (7.22).

CHAPTER 7

RECOMMENDATIONS

We make the following recommendations.

The number in ( ) indicates the paragraph in the main text where the recommendation is made.

The recommendations are grouped into the same topics as are used in the thematic summary.

Emissions and Their Impact

7.1 Further work should be done to identify the specific sources of carcinogenicity in diesel exhaust and the toxicological effects of defined fractions or combinations of components. (2.30)

7.2 The UK Government should propose that the European Community develops more discriminating techniques for analysing emissions from heavy duty diesel vehicles. This should be done in liaison with other authorities, with a view to introducing internationally agreed protocols on the measurement of such emissions. (2.34)

The Control of Emissions from New Vehicles and Engines

7.3 Consideration should be given to the possibility of consolidating the limit on smoke emissions into the Directive on other emissions from heavy duty diesel vehicles, 88/77/EEC. (3.3)

7.4 A more fundamental approach should be taken to setting emission limits for vehicles. Guide values for air quality should be set in the light of the best available data on the health effects and other impacts of the pollutants and should be reviewed periodically. (3.14) Measures designed to move towards achievement of those values should then be devised. These would include the setting of limit values for heavy duty and other vehicle emissions but would also need to address other factors. The Government should commence the necessary programme of work. (3.15)

7.5 Until the work described above has been done, we support the continuation of the present approach of aiming to reduce all emissions as far as is reasonably practicable. This is the most appropriate course of action in the light of present knowledge of the environmental impacts of the various pollutants. (3.16)

7.6 The findings of the expert committees with respect to the carcinogenicity of diesel particulates justify a precautionary approach which seeks to reduce such emissions from diesel engines as far as is practicable. There is also a strong case for continuing to seek reductions in emissions of NOx. On the basis of present knowledge, we consider that future emission limits, including those to be implemented in the European Community at the end of the decade, should require tighter control of both NOx and particulate matter; they should not concentrate on one to the exclusion of the other. This policy will need to be reviewed in the light of the further work on the impacts of emissions which we recommend above. It may also be influenced by technical developments. (3.17)

7.7   Consideration should be given to the setting of separate limit values for some of the various components of emissions from heavy duty diesel vehicles. (3.20)

7.8   We welcome the Government's initiative in proposing that an EC Directive on emissions from off-road diesel engines be prepared and recommend its speedy development. (3.55)

Engine Test Cycles

7.9   In considering the engine test to be used in the European Community for the next stage of limit values, to take effect at the end of the decade, the UK Government should press the European Community to develop one which uses equipment to the same specification as the US transient test but whose cycle emphasises the representation of European urban driving conditions. It should be defined well before the implementation date of the limit values to which it will apply. (3.32)

7.10   The UK Government should encourage the European Commission to raise the issue of testing heavy duty diesel engines in the UNECE, seeking wider international agreement to a harmonised procedure. The OECD and UNEP may also have a role to play in seeking the widest possible international agreement. (3.33)

7.11   The Government should press for the introduction of greater flexibility into the European Community's test procedure for heavy duty engines, along the lines of family certification. (3.38)

Economic Instruments

7.12   A practical form of incentive for the manufacture and the purchase of vehicles with less polluting engines would be differences in the rate of vehicle excise duty (VED). The rate of VED for a vehicle should depend, as well as on size and other factors, on whether its engine meets the emission values specified in successive stages of EC legislation. The rate categories could best be applied on the basis of the certificated performance of the engine. They should apply for the whole life of the vehicle, unless and until it is fitted with an engine which qualifies it for a different rate. (3.41)

7.13   The UK Government should press for EC Directive 88/77/EEC to be further amended to widen the scope for the application of economic instruments. In particular, the time for which incentives may be maintained should be extended and it should be permissible for them to be applied to vehicles with engines meeting the emission values specified in EC Directive 88/77/EEC or Stage I of the agreed amendment as well as to vehicles meeting Stage II. (3.42)

7.14   A grant should be offered for the fitting of a particulate trap to a new bus, unless and until the use of such a trap is required by the setting of tight new limits on particulate emissions from buses (recommendation 7.24). (3.48)

7.15   A grant should be paid towards the cost of a new engine in a new bus, meeting the emission values specified in one of successive stages of EC legislation, purchased before the relevant stage was implemented and the values became mandatory. (3.48)

7.16   If a package of retrofit measures enabled an engine to meet the emission values specified in a stage of EC legislation beyond the stage whose values it met previously, the vehicle should qualify for the appropriate lower rate of vehicle excise duty which we recommend at 7.12. (4.30)

7.17   Assuming that the trials recommended at 7.41 prove the feasibility of retrofitting traps, a grant at an appropriate level should then be offered for their retrofitting to any bus. (4.33)

7.18   A vehicle whose engine has been rebuilt to meet the emissions limit values specified in EC Directive 88/77/EEC, or in future the values in one of the stages of amendment, should qualify for the appropriate lower rate of vehicle excise duty recommended at 7.12. (4.37)

7.19   An additional incentive for engine rebuilding should be offered for buses in the form of a grant for the fitting of a rebuilt engine which meets the emission values specified in EC Directive 88/77/EEC, or in future the values in one of the stages of amendment, provided that the old engine is either scrapped or surrendered to be rebuilt to meet those values. (4.38)

7.20   A grant should be offered for replacing the old engine in a bus with a new one which meets tighter emission values. This would be on condition that the old engine were either scrapped or surrendered to be rebuilt to meet the emission values specified in EC Directive 88/77/EEC, or in future the values in one of the stages of amendment, for use in another vehicle. The size of the grant should reflect the certificated emissions performance of the replacement engine: one amount if the engine met whatever were the limits then in force and a larger one if it met limits which had been adopted but not yet implemented. (4.41)

7.21   The Government should introduce a fuel duty differential in favour of low sulphur fuel which is more than sufficient to off-set the additional cost of its production, thus compensating the producer and enabling the retail price to be lower. It should do so from the earliest possible date and should maintain the differential until the use of such fuel becomes mandatory. (5.8)

7.22   The Government should consider restricting the fuel duty rebate for buses to diesel fuel with a sulphur content not exceeding 0.05% at the earliest practicable date. (5.9)

7.23   Subject to its judgment of the study recommended at 7.50, the Government should seek to encourage the use of alternative fuels by buses and should consider the most appropriate means of doing so. One might be to offer a grant towards the cost of fuel storage and delivery equipment. Provided that the vehicle met the emission values specified in one of the stages of EC legislation pertaining to diesel vehicles, the appropriate rate of VED recommended at 7.12 should be charged. It should also attract the appropriate level of grant for a new bus or a replacement engine recommended at 7.15 and 7.20 respectively. (5.34)

The Urban Environment and Buses

7.24   Decisions on the emission limits to be set by the European Community at the end of the decade will need to be taken during the mid-1990s. It may by then be clear that, by the end of the decade, it will be practicable to fit particulate traps to all new heavy duty diesel vehicles. If not, however, and assuming that particulate emissions in urban areas continue to give rise to concern, consideration should be given to measures targeted on urban areas. These could include a much lower limit on particulate emissions from buses than from other heavy duty diesel vehicles, requiring buses to be fitted with particulate traps. (3.18)

7.25 By the mid-1990s it may be clear that, by the end of the decade, it will be practicable to fit flow-through catalysts to all new heavy duty diesel vehicles to control emissions of NOx. If not, however, consideration should be given to setting a much lower limit on NOx emissions from buses than from other heavy duty diesel vehicles, requiring buses to be fitted with such catalysts. (3.19)

7.26 Many of the considerations which apply to buses apply also, to varying extents, to other classes of vehicle which operate mainly in urban areas. The Government should carry out the necessary work to identify the classes of vehicle which make the largest contributions to emissions in urban areas and should consider how our recommendations for buses might be applied to them. (1.11c)

Recommendations 7.14, 7.15, 7.17, 7.19, 7.20, 7. 22, 7.23, 7.37, and 7.41 are also relevant to buses.

Standards and Tests for Vehicles in Service

7.27 The US authorities require that engines be designed not to deteriorate in emissions performance by more than a specified factor over prolonged periods of operation. Such an approach should be considered by the UK Government for incorporation into the European Community's engine certification procedures. (4.5)

7.28 The practicability of requiring that appropriate maintenance be carried out on heavy duty diesel engines should be considered. (4.6)

7.29 It will shortly be possible to require that an engine should be kept in good condition as defined by an appropriate set of parameters. The Government should develop this approach as part of its own legislative control of emissions from heavy duty diesel vehicles in service and should propose its introduction throughout the European Community. (4.7)

7.30 The Government, in liaison with the relevant trade associations, should ensure that the standards of training and qualification for those maintaining heavy duty vehicles are reviewed and should initiate any action needed to ensure that the necessary standards are achieved by all operators. (4.8)

7.31 We welcome the Government's intention to make the test for smoke emissions from heavy duty vehicles in service more severe and urge the Government to introduce the change as soon as possible. (4.10)

7.32 The UK Government should consider introducing, as part of the smoke test for heavy duty vehicles in this country, a requirement that the vehicle be run on rollers against a load applied by its own brake. (4.11)

7.33 Development work should be carried out to determine the range of heavy duty engines to which diagnostic testing may be applied and to extend that range as far as is practicable. A test of engine condition should be added to the annual test for heavy duty vehicles in this country. (4.12)

7.34 Research should be carried out into the application of advanced engine diagnostic techniques to emissions control. The Government should sponsor such research if necessary. It should also consider the feasibility of requiring the incorporation of the necessary sensors and other equipment into new heavy duty engines. (4.13)

7.35 The UK Government should press for the introduction, throughout the European Community, of diagnostic engine testing and possibly for the introduction of a 'roller' test for smoke. (4.14)

7.36 The Vehicle Inspectorate should carry out more spot checks of emissions from heavy duty vehicles; we welcome the Government's commitment in the 1990 environment White Paper to do so. The checks should incorporate the new instrumented measure of smoke and diagnostic techniques to indicate performance in controlling other emissions. (4.17)

7.37 We welcome the statement that the Government will place considerable emphasis on the need to maintain good emissions performance when licensing heavy goods vehicle operators. Similar emphasis should be placed on emissions performance when licensing operators of passenger service vehicles. (4.18)

7.38 The Vehicle Inspectorate should seek to extend the scope of its annual smoke survey by investigating new techniques for the remote sensing of a range of diesel emissions. (4.20)

7.39 Smoky vehicle spotter programmes should be developed, with or without public involvement. More local authorities should consider introducing them and the Vehicle Inspectorate, the Police and the Driver and Vehicle Licensing Centre should cooperate fully with them. (4.23)

7.40 We commend the Government's recent initiatives in encouraging members of the public to report smoky vehicles direct to the Vehicle Inspectorate. Permanent arrangements should be made to ensure that the public is aware of how to report smoky vehicles to the Inspectorate. (4.24)

7.41 The Government should proceed urgently with trials of particulate traps, concentrating on their application to buses. A substantial grant towards the cost should be offered for any vehicle taking part in the trial. (4.33)

Fuels and Lubricants

7.42 The UK Government should consider introducing legal control of the cetane number and perhaps other characteristics of diesel fuel. It should also consider proposing that an EC Directive on this be introduced. (5.15)

7.43 The UK Government and the European Commission should commission research into the relationship between defined aromatic compounds in diesel fuel and the composition of emissions and to assess the environmental hazards which the emissions present. (5.18)

7.44 If it is shown that specific aromatic components in diesel fuel result in harmful emissions, the concentration of those components in the fuel should be limited. (5.18)

7.45 The Government should use its new powers for controlling the use of fuel additives to the full. It should adopt the following approach:-

The use of potentially hazardous metals as fuel additives should be banned until the combustion products of such additives, emitted from the exhaust, have been subjected to appropriate toxicological testing.

No new substance should be permitted to be used as a fuel additive until similar testing has been carried out on it.

A programme of such testing on existing substances used as fuel additives should be established as a basis for reviewing their continued use. (5.23)

7.46   If a metal additive is used for regeneration of a particulate trap and most of the metal is retained within the filter element, this should be taken into account in evaluating the environmental impact of the emissions. If the metal is potentially hazardous, care should be taken in the eventual disposal of the filter. (5.23)

7.47   The merits of introducing a further improved diesel fuel should be kept under review, taking account of the energy cost of its production. (5.24) If one is introduced, its use should be required from an appropriate date. In the interim, financial incentives should be created for the use of the new fuel similar to those we recommend for 0.05% sulphur fuel at 7.21 and 7.22. (5.25)

7.48   When more discriminating techniques for measuring emissions have been developed, and more selective limit values adopted, the Government should consider whether the development of a diesel fuel standard related to emissions performance is then feasible and desirable. (5.27)

7.49   Investigation of an alternative basis for a standard for lubricants, related to emissions performance, should be undertaken. (5.29)

7.50   The Government should conduct a study of the effect of the production and use of alternative fuels on carbon dioxide and other emissions. (5.34)

Recommendations 7.21, 7.22 and 7.23 are also relevant to fuels and lubricants.

Acknowledgement

We are grateful to our consultant David Broome whose expertise guided us through many complexities. Many individuals and organisations, listed in Appendix 2, contributed in other ways to our study; to them too we offer thanks. We also express appreciation of our secretariat, particularly Philip Dale and Richard Wakeford, who supported the study with skill and dedication.

ALL OF WHICH WE HUMBLY SUBMIT FOR YOUR MAJESTY'S GRACIOUS CONSIDERATION

Lewis (Chairman)
Cranbrook
Barbara Clayton
Henry Charnock
Henry Fell
Peter Jacques
John Lawton
J Gareth Morris
Jeremy Pope
Donald Reeve
Emma Rothschild
William Scott
Aubrey Silberston
Charles Suckling

B Glicksman   Secretary

P S Dale   Assistant Secretary

September 1991

APPENDIX 1

Members of the Royal Commission and Consultant for the Study

Chairman

THE RT HON THE LORD LEWIS OF NEWNHAM, Kt, MA, MSc, PhD, DSc, ScD, CChem, FRSC, FRS
> Professor of Inorganic Chemistry, University of Cambridge
> Warden of Robinson College, Cambridge

Members

PROFESSOR H CHARNOCK, MSc, DIC, FRS
> Emeritus Professor of Physical Oceanography, University of Southampton
> Chairman Meteorological Research Sub-committee, Meteorological Committee

PROFESSOR DAME BARBARA CLAYTON, DBE, MD, PhD, HonDSc (Edin), FRCP, FRCPE, FRCPath
> Honorary Research Professor in Metabolism, University of Southampton
> Past-President, Royal College of Pathologists
> Chairman, MRC Committee on Toxic Hazards in the Environment and the Workplace
> Deputy Chairman, Department of Health Committee on Toxicity of Chemicals in Food, Consumer Products and the Environment
> Chairman, Standing Committee on Postgraduate Medical Education
> Honorary Member, British Paediatric Association

THE RT HON THE EARL OF CRANBROOK, MA, PhD, DSc, DL, FLS, FIBiol
> Partner, family farming business in Suffolk
> Chairman, Nature Conservancy Council for England
> Chairman, Institute for European Environmental Policy [London]
> Non-executive Director, Anglian Water plc
> Member, Broads Authority and Harwich Haven Authority
> Vice-President, National Society for Clean Air and Environmental Protection

MR H R FELL, FRAgS, NDA, MRAC
> Managing Director, H R Fell and Sons Ltd
> Council Member, Royal Agricultural Society of England
> Member, Minister of Agriculture's Advisory Council on Agriculture and Horticulture (1972-81)
> Commissioner, Meat and Livestock Commission (1969-78)
> Past-Chairman, The Tenant Farmers Association

MR P R A JACQUES, CBE, BSc
> Head, TUC Social Insurance and Industrial Welfare Department
> Secretary, TUC Social Insurance and Industrial Welfare Committee
> Secretary, TUC Health Services Committee
> Secretary, TUC Pensioners Committee
> TUC Representative, Health and Safety Commission
> TUC Representative, Social Security Advisory Committee

PROFESSOR J H LAWTON BSc, PhD, FRS
  Director, Natural Environment Research Council Interdisciplinary
    Research Centre for Population Biology, Imperial College,
    Silwood Park
  Professor of Community Ecology, Imperial College of Science,
    Technology and Medicine
  Member, British Ecological Society
  Member, American Society of Naturalists
  Council Member, Royal Society for the Protection of Birds

PROFESSOR J G MORRIS BSc, DPhil, FIBiol, FRS
  Professor of Microbiology, The University College of Wales,
    Aberystwyth
  Chairman, SERC Biological Sciences Committee (1978-1981)
  Chairman, UGC Biological Sciences Committee (1981-1986)
  Member, Society for General Microbiology

MR J J R POPE, OBE, MA, FRSA
  Deputy Chairman and Managing Director, Eldridge, Pope and Co plc,
    Brewers and Wine Merchants
  Chairman, The Winterbourne Hospital plc (1981-1989)
  Deputy President, Food and Drinks Federation (1987-1990)
  Member, Top Salaries Review Body

MR D A D REEVE, CBE, BSc, FEng, FICE, FIWEM
  Deputy Chairman and Chief Executive, Severn Trent Water Authority
    (1983-85)
  Past-President, Institute of Water Pollution Control
  Past-President, Institution of Civil Engineers
  Member, Advisory Council on Research and Development,
    Department of Energy

EMMA ROTHSCHILD, MA
  Senior Research Fellow, King's College, Cambridge
  Research Fellow, Sloan School of Management, Massachusetts
    Institute of Technology (MIT)
  Associate Professor of Science, Technology and Society, MIT (1978-
    1988)
  Member, OECD Group of Experts on Science and Technology in the
    New Socio-Economic Context (1976-1980)
  Board Member, Stockholm Environment Institute

MR W N SCOTT, OBE, BSc, FIChemE, FInstPet, FInstD
  Director, Shell International (1977-85)
  Non-executive Director, Anglo and Overseas Investment Trust
  Non-executive Director, Shell Pension Trust
  Consultant, UK and Japanese companies
  Past-Chairman, CONCAWE

PROFESSOR Z A SILBERSTON, CBE, MA
  Senior Research Fellow, Management School, Imperial College of
    Science, Technology and Medicine
  Professor Emeritus of Economics, University of London
  Secretary-General, Royal Economic Society
  Member, Restrictive Practices Court
  Past President, Confederation of European Economic Associations

Dr C W Suckling, CBE, PhD, DSc, DUniv, CChem, FRSC, Senior
Fellow RCA, FRS
General Manager for Research and Technology, Imperial Chemical
Industries (1977-82)
Consultant in science, technology and innovation
Honorary Visiting Professor, University of Stirling

Consultant
Mr D Broome, MA, CEng, FIMechE, MIRTE, MSAE
Director, Ricardo Consulting Engineers Ltd.

APPENDIX 2

Organisations and Individuals Contributing to the Study

Listed below are those organisations and individuals who gave written evidence or assisted the Commission in other ways during the study. Those marked* gave oral evidence or a factual presentation at a meeting of the group of members of the Commission who took the study forward (see Preface). Those marked+ gave oral evidence during visits by the group, details of which are listed at the end of this appendix.

Government Departments
HM Customs & Excise
Department of Energy
Department of the Environment*
Department of Health
Department of Trade and Industry: Warren Spring Laboratory
Department of Transport* including
    Transport and Road Research Laboratory
    Vehicle Inspectorate Executive Agency
Foreign & Commonwealth Office: Embassy in Tokyo
Health & Safety Commission and Executive
Medical Research Council: Environmental Epidemiology Unit
Scottish Development Department

Other Organisations
Association of Chief Police Officers
Blackburn Borough Council
BP Oil, USA
British Railways Board
Bus and Coach Council
Chloride Silent Power Ltd
Commission of the European Communities+
Committee of Common Market Automobile Constructors
CONCAWE, the oil companies' European organization for environmental
    and health protection
Convention of Scottish Local Authorities
Council for the Protection of Rural England
Cummins Engine Co Ltd
DAF BV
Degussa AG
Derby City Council
ECOTEC
Engine Manufacturers Association, USA
Environmental Protection Agency, USA
Fellowship of Engineering
Freight Transport Association
Friends of the Earth
ICI Chemicals & Polymers Ltd
Institute of Petroleum
Institute of Road Transport Engineers
Institution of Environmental Health Officers
IVECO FIAT SpA
Johnson Matthey Catalytic Systems

Kirklees Metropolitan Council
London Boroughs Association
Lubrizol Ltd
Lucas Diesel Systems
MAN Nutzfahrzeuge AG
Mercedes-Benz AG
Motor Industry Research Association
National Society for Clean Air and Environmental Protection
PARAMINS, Exxon Chemical Technology Centre
Perkins Technology Ltd+
Renault Vehicules Industriels
Ricardo Consulting Engineers Ltd
Road Haulage Association
Robens Institute of Health and Safety
SAAB SCANIA A-B
Shell Internationale Petroleum Mij
Shell UK Ltd
Society of Motor Manufacturers and Traders
Technical Committee of Petroleum Additive Manufacturers in Europe
  (ATC)*
Transport and Environment Studies (TEST)
Umweltbundesamt (Federal Environmental Agency, Germany)+
UK Petroleum Industries Association*
University of Leeds
Volvo Truck Corporation
Volvo Trucks (GB) Ltd
West Glamorgan County Council
World Wide Fund for Nature UK

Individuals

Dr D J Ball	University of East Anglia
Mr J M Dunne*	Warren Spring Laboratory
Dr M F Fox	Leicester Polytechnic
Dr M Green	British Lung Foundation
Mr J A Terning	CONCAWE

Visits

During the course of the study members of the Commission visited the
organisations listed below:

4 December 1990:	Perkins Technology Ltd, Peterborough
8 April 1991:	Commission of the European Communities, Brussels
15 April 1991:	Umweltbundesamt, Berlin

APPENDIX 3

The Commission's Invitation for Evidence, July 1990

The Royal Commission has embarked upon a study on environmental pollution arising from emissions from heavy duty diesel vehicles. The study will be concerned with the emission of gases and particulates (including smoke), focusing on:

> future emission standards, with particular reference to the proposed amendment to EC Directive 88/77/EEC;

> means of abatement of emissions, including the scope for retrofitting vehicles in service;

> the development and enforcement of standards to be met by vehicles in service;

> quality standards for diesel fuel.

The study will take account of the environmental impact of exhaust and other emissions, the scope for their abatement, the pattern of vehicle use and the financial implications of any measures proposed.

Your organisation is invited to submit evidence for this study. It forms part of a new series of short studies which the Commission will undertake. The Commission's aim is to complete the study early in 1991. It would therefore be helpful if your written evidence could be submitted in two parts:-

> Any relevant papers which are published or readily available. I should be grateful to receive these by the end of this month if possible, together with any suggestion for other sources of information.

> The remainder of your evidence, including your organisation's views on some or all of the aspects of the topic identified above. It would be helpful if this could reach me by 14 September.

The Commission would welcome evidence on those aspects of the topic which fall within your organization's field of expertise. You are of course free, if you wish, to comment on other aspects, not listed above, which directly or indirectly have an impact on the environment. Since this study is intended to be a short one, however, it may be necessary for the Commission to limit the scope of its Report.

Signed by the Secretary to the Commission

APPENDIX 4

DIESEL ENGINE EMISSIONS CONTROL TECHNOLOGY

A review prepared in 1991 by Ricardo Consulting Engineers Ltd at the request of the Royal Commission on Environmental Pollution. The views expressed here are those of Ricardo Consulting Engineers Ltd and do not necessarily reflect the views of the Royal Commission. Several of the figures referred to in the text are reproduced in Chapter 2 of the Report.

Contents

1. Introduction

The diesel engine has established for itself a position of total dominance as a heavy duty, mobile power source in both on-road (truck and bus) and off-road (agricultural and construction equipment) applications. Several attributes account for this dominance, but the major features can be simplified to good durability and good fuel efficiency, at low cost per horsepower.

The heavy duty diesel engine has traditionally been seen by the layman as noisy, with a tendency to emit black smoke. The engineer and the environmentalist understand the diesel engine as being fundamentally cleaner than the petrol engine, emitting lower quantities of the exhaust pollutants currently legislated against, and significantly less carbon dioxide, of growing concern with regard to the greenhouse effect.

Weighing against the diesel engine are concerns over its emissions of oxides of nitrogen (NOx) and the potential health effects from particulate matter exhausted into the atmosphere. For this reason, legislation to control both NOx and particulate output is either current or planned for most major markets throughout the world.

In order to follow any discussion relating to the control of diesel exhaust emissions it is necessary to have an elementary understanding of the workings of the diesel engine, and the processes leading to the formation of the various pollutants.

2. Heavy Duty Diesel Engine Operation

The thermodynamic cycle championed by and named after Rudolph Diesel has been universally exploited in the form of reciprocating piston engines although details of the method of exploitation have varied. Within the world of mobile, heavy duty applications the four-stroke, direct injection diesel engine has at present and is expected to retain close to total dominance. Throughout this paper it may be assumed that, except where stated otherwise, it is this type of diesel engine which is being described or discussed.

2.1 *Working Principles*

The key points of the working of a 4-Stroke Engine are shown in Figure 2.1 (page 7).

1st (Inlet) Stroke — The working principles can best be described starting with the piston travelling downwards and the inlet valve (or valves) open. During this stroke air is drawn into the engine. Just after the bottom of the piston travel the inlet valve (or valves) close, trapping a cylinder full of air. Often (but not always) the air inlet system can be shaped to force the air charge to rotate (as with water leaving a bath) as it enters the engine, this swirling continuing during the following stroke and assisting the later mixing of air and fuel.

2nd (Compression) Stroke — With the inlet and exhaust valves closed the piston travels upwards, compressing the trapped air charge into a cavity in the top of the piston, the final volume of the air being typically 6% of the original volume. As a result of this compression there is a rise in temperature (in excess of 500°C) and pressure (in excess of 45 times the inlet pressure). To this charge of air at high temperature and pressure is now added fuel. The fuel is pumped up to high pressure, and injected into the engine through several small holes in the injector nozzle tip, such that finely atomised fuel is distributed throughout the hot pressurised air.

3rd (Expansion) Stroke - After a short delay (typically 0.5–1 milliseconds) the injected fuel begins to burn. This burning then continues as relative movement between the fuel droplets and vapour and the air brings the fuel into contact with fresh oxygen, until either all of the fuel or all of the available oxygen is consumed. During the burning phase heat energy is released, increasing the pressure of the cylinder gases. This pushes the piston back downwards, delivering the required power to the crankshaft.

4th (Exhaust) Stroke — Close to the bottom of the piston travel the exhaust valve (or valves) open. The rotational inertia of the engine causes the piston to rise, forcing the burnt exhaust gases out through the engine exhaust system.

This cycle has required two revolutions of the engine and can now repeat from the first stroke.

From the brief description above it will be understood that the availability of an adequate air supply is fundamental to successful full load operation of the diesel engine. Methods of maximising air supply are discussed in section 4 of this paper.

3. Sources of Exhaust Emissions

The diesel engine operates always with more air available than the theoretical minimum requirement for complete combustion of the fuel. The optimum requirement is for the hydrocarbon fuel (Hydrogen and Carbon) to burn in air (Oxygen and Nitrogen), exhausting carbon dioxide (CO_2) water (H_2O), nitrogen (N_2) and unused air. None of these emissions are limited by legislation, although concern is growing over the contribution of global CO_2 emissions to the greenhouse effect.

All the hydrocarbon fuel burning engines produce, and indeed have to produce, CO_2. No scope therefore exists to reduce this particular emission *per se*, except by burning less fuel. The inherently superior fuel economy of the direct injection diesel engine over other prime movers thus makes it a fundamentally lower producer of CO_2.

3.1 Emissions Addressed by Legislation

In practice the combustion which occurs in real engines (petrol or diesel) is not as simple as the 'perfect' combustion described above. Other constituents present in the fuel, incomplete combustion and undesired reactions result in further pollutants being exhausted. Several of these have been identified as present in sufficient quantity to be of environmental or health concern. Legislation to control these pollutants was first introduced in the USA in the early 1970s for road going vehicles. This has been followed by similar legislation in most major markets around the world, and will before long be adopted for off-road applications, (eg agricultural and construction equipment) in several countries.

The emissions which are subject to legislation and their sources in the heavy duty diesel engine are discussed below.

3.1.1 Carbon Monoxide

It has previously been stated that, where sufficient air is available, full oxidation of the carbon constituent of the hydrocarbon fuel should be possible (ie to carbon dioxide, CO_2). In practice, the short time available for the reaction, and the failure of some of the fuel to mix and react with sufficient oxygen results in incomplete oxidation to carbon monoxide (CO) which is odourless and toxic to human beings.

The legislation to control CO emissions has been framed to control emissions from petrol engines, which, by their fundamental mode of combustion are high producers of CO unless complex control methods are employed. The mode of operation of the diesel engine is that a full charge of air is drawn in for every cycle and the amount of fuel injected is reduced for part load operation. An excess of air is therefore generally available and the formation of CO is inhibited. Carbon monoxide levels from diesel engines are always well within any limit set for petrol engines, and are not usually of concern to the diesel engineer.

Main Control Methods:

— Carbon monoxide should not need controlling for diesel engines.

3.1.2 *Unburnt Hydrocarbons*

Viewed simply it can be noted that some of the hydrocarbon fuel fails to burn in the combustion chamber. The hydrocarbons exhausted to atmosphere react in the presence of oxides of nitrogen and sunlight to form oxidants (visible as a smog on sunny days in cities with heavy traffic). Several mechanisms can contribute to unburnt hydrocarbon emissions, but two are of constant concern to the diesel engineer:-

Mode of Formation — Under certain operating modes the conditions in the cylinder are not conducive to complete combustion. During start-up from cold, the metal temperatures within the combustion system will reduce the compression temperature of the compressed air charge. This not only inhibits the start of the combustion process but also hinders successful completion by prematurely quenching burning fuel droplets. Furthermore, any of the fuel spray which becomes deposited on the cold internal metal surfaces will not evaporate off and will pass through the system unburnt. When starting from cold, the level of these unburnt hydrocarbons can be excessive, and are evident as 'white smoke' from the exhaust system. Once the engine has reached operating temperature, other modes of operation have a tendency to cause emission of high levels of unburnt hydrocarbons.

When operating at high rotational speeds the time available for the combustion process to complete is short. Under part load conditions (ie when fuelling levels are reduced) fuel/air mixing problems may result in late and/or incomplete combustion. These factors both tend to lead to increased emission of unburnt hydrocarbons and the worst cases (poorly developed or maintained engine, poor or incorrect fuel) excessive levels can result from conditions of mis-fire.

With an engine operating at idle (the lowest speed and load condition), internal component temperatures decrease and mixing problems associated with the very small injected fuel quantity inhibit good combustion, causing increased output of unburnt hydrocarbons.

Fuel Injection Equipment — The design and optimisation of the fuel injection equipment can be a major contributor to controlling emissions of unburnt hydrocarbons. The major considerations are:-

Atomisation — Fuel injection at low pressures through nozzles with large diameter holes results in poor fuel atomisation. These large fuel droplets expose less total surface area for evaporation and burning than a greater quantity of smaller droplets. Furthermore the body of fuel within the droplets has a tendency towards pyrolysis. These factors tend to increase unburnt hydrocarbons emissions.

Design — The nozzles fitted to conventional diesel fuel injectors contain a small volume of fuel downstream of the control needle, contained in the nozzle sac and within the holes (Figure A4.1). A proportion of this fuel will tend to enter the combustion process late in the cycle, after the main injection is complete. Failing to burn fully, it will enter the exhaust as unburnt hydrocarbon.

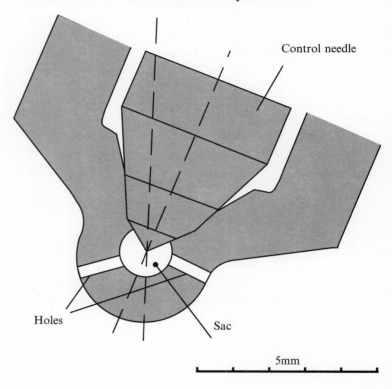

Figure A4.1 Diesel fuel injector nozzle
The tip of the nozzle with its sac and holes. The flow of fuel is controlled by the needle. The nozzle, whose total length would typically be about 35mm, is inserted into the cylinder at an angle as shown.

Other features relating to the optimisation of the fuel pump and injector design, which result in fuel entering the combustion system at low pressure or low rate, late in the cycle, will tend to contribute to levels of unburnt hydrocarbon emissions.

The above explanation of sources of unburnt hydrocarbons has assumed simplistically that the sole source is fuel, as input, passing straight through the system. In practice both fuel and lubricating oil reaching the combustion space past the piston rings contribute to emissions of hydrocarbons which may be unburnt, partially burned or thermally cracked. Analysis of emitted hydrocarbons therefore shows a complicated combination of quantities and species.

Main Control Methods:

— advanced injection timing;

— increased compression ratio;

— improved fuel injection equipment;

— increased air supply;

— oxidisation with a catalyst in the exhaust (discussed in section 5).

82

3.1.3 *Oxides of Nitrogen*

The nitrogen present in air undergoes an oxidation reaction of its own during combustion, being emitted by the engine as nitric oxide (NO) and nitrogen dioxide (NO_2). These two compounds are customarily treated together as NOx. NO rapidly oxidises to NO_2 once exhausted from the engine. The level of NOx produced depends upon the temperatures achieved during the reaction and on the length of time spent at these temperatures.

The combustion conditions to produce low levels of NOx in the exhaust are generally not those favouring low levels of hydrocarbons. High temperatures and long reaction times producing high levels of NOx will tend to ensure effective combustion of hydrocarbons. Conditions favouring low NOx formation, ie lower combustion temperatures and shorter reaction times, are furthermore generally in opposition to the requirements for low smoke, good fuel economy and low exhaust temperatures. The diesel engineer is thus faced with the paradox that the NOx levels which need to be reduced are in general a sign of 'good' combustion.

Main Control Methods:

— retarded injection timing;

— reduced combustion temperatures;

3.1.4 *Black Smoke*

It has previously been explained that during each cycle the diesel engine draws in a full charge of air. The work output required from the engine is then set by controlling the quantity of fuel delivered by the fuel injection system. Under light load operation there is thus more air available than the minimum requirement for complete combustion of the fuel supplied. However, as the load on the engine is increased the ratio of air to fuel decreases, approaching the minimum requirement. As this limit is approached it becomes more difficult for localised droplets of fuel to meet and successfully mix with sufficient air in the time available (say of the order of 5 milliseconds), resulting in partial combustion to carbon (soot). In sufficient quantities these soot particles become visible in the exhaust stream as black smoke. Traditionally it has been the level of visible smoke which is considered acceptable which has set the limit on maximum engine output.

Main Control Methods:

— increased air supply;

— reduced rating with the consumption of less fuel;

— reduced friction;

— improved mixing and combustion.

3.1.5 *Particulate Matter*

Legislation with regard to particulates from heavy duty diesels has been in force in the USA since 1988 (light duty since 1984) and is now being introduced, or considered for introduction, in most major markets worldwide. This pollutant is in practice specific to the diesel engine, due to the inherent heterogeneous nature of the combustion and the composition of the fuel needed to achieve satisfactory ignition by charge compression. Diesel particulate is not simply definable in chemical terms, its definition being "that matter collected on a filter paper" under certain stated conditions. The measuring conditions require that the raw exhaust from the engine be diluted with air under controlled conditions, allowing the exhaust constituents to cool and allowing processes of condensation, agglomeration and adsorption to take place prior to filtration. The objective of this procedure is to simulate the conditions of diesel exhaust leaving a vehicle. The filter paper specification is selected such that the matter collected represents respirable particles.

The constituents of diesel particulate are many and varied, and have been the subject of many studies. For the purposes of this paper they will be simplified as follows:-

Insoluble Organic Matter – this is principally carbon resulting from the processes described above for black smoke, present even when smoke is not visible to the naked eye. It is insoluble in organic solvents.

Soluble Organic Matter – principally hydrocarbons collected either in free form or adsorbed onto the surfaces of the insoluble matter. The sources of the hydrocarbons are both in fuel and lubricating oil.

Remainder – products arising from the sulphur content of the fuel, condensed water vapour and trace constituents of the fuel and ash which may be inorganic.

In simple terms the sources of exhaust particulate matter from diesel exhausts can be considered primarily as those responsible for the unburnt hydrocarbons and black smoke. However sulphur present naturally in conventional diesel fuel will be converted to sulphate, some of which will be trapped as particulate.

Main Control Methods:

— increased air supply to reduce smoke;

— reduced rating with the consumption of less fuel to reduce smoke;

— improved fuel injection to reduce hydrocarbon components;

— a catalyst fitted in the exhaust system to reduce hydrocarbons components only;

— reduced fuel consumption;

— reduced oil consumption;

— advanced injection timing;

— reduced fuel sulphur content;

— maintenance of fuel ignition quality;

— a particulate trap in the exhaust system.

4. Primary Methods of Emissions Control

Consideration of various factors (in particular cost, complexity, durability etc) dictates that efforts should be directed at reducing, as far as possible, the formation of the pollutants at source. Several techniques and design guidelines have become accepted as standard practice for diesel engines being developed for use in applications where exhaust emissions are now, or are expected soon to be, controlled.

4.1 *Air Supply*

Whereas light duty (passenger car) engines rarely operate in service close to their maximum output potential for any significant time, heavy duty engines (as used in trucks, tractors, excavators etc) are major working power sources and are expected to spend significant operating periods working at, or close to, their maximum potential.

It has previously been explained that the output of the diesel engine is increased simply by injecting more fuel, up to the point when insufficient air remains for complete combustion. At that point levels of black smoke, unburnt hydrocarbons and particulate matter present in the exhaust will have risen to excessive levels. In order to reduce these emissions, and to increase power levels, it is necessary to increase the air supply to the engine.

4.1.1 *Breathing Improvement*

During each cycle, an engine fills the cylinder with a volume of air equal to the swept volume of the cylinder (bore area x stroke). However, as air is drawn into the cylinder aerodynamic losses in the intake system cause a pressure drop and heat transfer from the hot metal components and the effects of work on the air charge cause a temperature rise. The result of this is that the volume of air drawn in from the atmosphere is less than the swept volume of the cylinder, the ratio of these volumes being the volumetric efficiency of the system. For a naturally aspirated engine (one in which the air supply is not pressurised) the volumetric efficiency is typically as low as 80% for a poorly optimised system, or as high as 90% if the intake system (intake ducts, valve timing, etc) are well designed and optimised. Further gains can be made if use is made of the dynamic effects of pressure pulses, originating from the opening and closing of the inlet valves, which travel along the intake ducts. Such tuned (or ram) systems can increase volumetric efficiency to 100% or more, but are often not appropriate for use, depending on other factors such as engine speed range, installation problems etc.

Improvements to the breathing capability of the engine will always enable exhaust emission levels to be reduced, either directly as with smoke or because more freedom exists for other control techniques without loss of performance. However the improvements available by this method are limited and militate against the naturally aspirated diesel engine where very low levels of exhaust emissions are required, as now targeted in Europe for the middle of the decade.

4.1.2 *Turbocharging*

Up to one third of the fuel energy supplied to a typical diesel engine is rejected as a loss through the exhaust system. The exhaust driven turbocharger recovers a proportion of this energy by expanding the hot exhaust gas through a turbine. This turbine in turn drives a compressor which raises the pressure (and hence the density) of the air charge supplied to the engine. A typical pressure ratio for a commercial heavy duty diesel engine is around 2:1.

With perhaps twice the mass of air available for combustion it is possible to increase significantly the quantity of fuel supplied (thus raising the power output from the engine) while retaining a greater quantity of excess air than was available to the naturally aspirated engine. In raising the pressure of the air charge, the turbocharger also causes a rise in the latter's temperature. The magnitude of this temperature rise will depend on the pressure increase and the efficiency of the compressor; for a typical turbocharged engine in a very large vehicle a delivery temperature in excess of 150°C may be expected.

It may now be summarised that the excess air available to the turbocharged diesel engine will enable a reduction in black smoke, unburnt hydrocarbons and particulates. However the increase in air charge temperature will result in an increase in NOx production.

4.1.3 *Aftercooling*

The heated air delivered by the turbocharger may be cooled by passing it through a heat exchanger, rejecting the heat to any convenient lower temperature medium. This in practice is usually either the engine cooling water (which will have been cooled by the radiator to 80-90°C) or ambient air. Cooling the charge air further increases the air density (either increasing the mass of air delivered or reducing the pressure required) and enables the above discussed improvement to smoke, hydrocarbon and particulate emissions to be achieved without an increase in NOx levels.

4.1.4 *Air Motion*

Before successful combustion can occur successful mixing must take place between the atomised fuel and the available air. Where fuel injection pressures are low and fuel droplets relatively large it is necessary to encourage mixing by means of high levels of air motion. This is usually achieved primarily by imparting a rotating or 'swirling' motion to the incoming air charge.

Where higher fuel injection pressures are adopted the fuel becomes better atomised and is more able to successfully mix with the available air. This enables lower levels of air swirl to be adopted, improving fuel efficiency as the heat transfer between the now slower moving combustion gases and the internal surfaces of the combustion chamber are reduced.

4.2 *Fuel Supply*

The design of the equipment supplying high pressure fuel to the diesel engine, and the characteristics of that fuel supply, are of paramount importance where optimum performance together with minimised emissions are sought. The key objectives are to produce well atomised fuel sprays injected at the optimum time, in the optimum place (direction), and at the optimum rate for thorough mixing with the air in the cylinder.

The fundamental features which must be addressed are:

- maximum injection pressure;
- injection rate characteristics;
- fuel injector and nozzle design;
- fuel spray characteristics;
- timing of injection.

4.2.1 *Maximum Injection Pressure*

A major requirement in order to achieve successful mixing of the fuel and air (a prerequisite for complete combustion with minimum emission of smoke, hydrocarbons and particulates) is that the fuel be finely atomised. This requires that the size of the holes in the injection nozzle must be kept as small as possible. However for the thermodynamic cycle to be optimised, the time taken to inject the required amount of fuel must be kept suitably short, necessitating high injection pressures. It will be apparent therefore that the higher the pressure capability of the fuel injection pump, the smaller can be the holes through which the fuel is injected.

Historical evidence shows how the diesel engineer has always sought to use higher pressures, and has pushed the fuel injection equipment manufacturers to increase the capability of their products. Early, unsophisticated diesel engines were operated with maximum fuel injection pressures of 200 bar or less. As improvements in fuel economy were sought, these pressures were increased to above 400 bar. The requirement for intermediate emission levels required that injection pressures were increased to around 800 bar, while the latest generation of low emission engines entering production are utilising 1000 to 1500 bar.

4.2.2 *Injection Rate*

It should be noted that during each injection event the fuel injection pump pressurises a metered quantity of fuel. The injection process starts at minimum pressure, increases to the maximum capability of the fuel pump, then drops back to the minimum. By careful design of the total fuel injection system the rate of injection at the start and the end of the process can be tailored to give the required characteristics for minimising exhaust emissions.

The required characteristics can best be understood by considering how the energy is released from the fuel during the combustion process (Figure A4.2).

As fuel starts to enter the hot compressed air charge in the combustion system at the start of the injection process it does not start to burn immediately. A short period elapses (called the delay period, length dependent on physical factors but also the quality of the fuel) during which the fuel droplets can vaporise and mix with the air. Once combustion commences this pre-mixed fuel burns rapidly causing a sharp rise in pressure and temperature in the cylinder. It will be remembered that these are the conditions

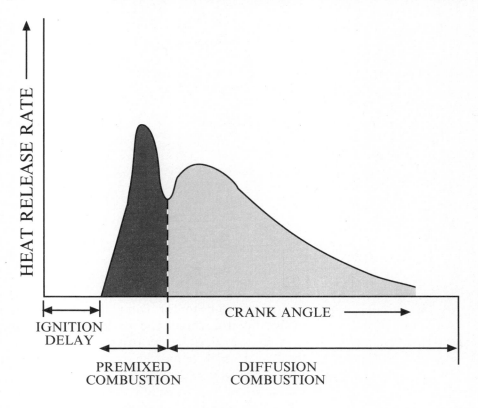

Figure A4.2 Heat release characteristics of a diesel engine

which lead to the formation of oxides of nitrogen. The importance of limiting the initial rate of injection to control NOx levels can thus be understood.

After the initial period of pre-mixed burning, combustion enters the main phase of diffusion burning. During this phase the important features are the minimising of droplet size and maximising of fuel and air mixing as discussed above.

The final phase of combustion occurs as the fuel injection pressure drops and injection nears completion. During this phase it is important to maintain a high injection pressure as long as possible, and then to cut off the injection as sharply as possible. It is known that failure to achieve this results in low pressure, poorly atomised fuel entering the combustion system late in the cycle. This fuel does not achieve full mixing and leads to the formation of carbon, unburnt hydrocarbons and particulate matter.

4.2.3 Fuel Injector and Nozzle Design

Several features in the design of fuel injectors contribute to the control of exhaust emissions:

— reduction in volume of fuel in sac and holes – discussed earlier with regard to hydrocarbon formation (Figure A4.1);

— reduction in mass/inertia of moving parts – assists in obtaining rapid termination of injection with benefits to smoke and particulate levels;

— increasing application of two-spring injectors – allows early injection of a small quantity of fuel at low rate prior to main injection, benefiting NOx control as described above;

— reduction in overall injector size – assists in optimum positioning of injector with benefits to be discussed later.

4.2.4 Fuel Spray Characteristics

A further feature which is fundamental to obtaining the optimum mixing of fuel and air is the number of holes in the injection nozzle and their geometric positioning. If incorrectly designed or not fully optimised during development of the engine this feature will lead to increased levels of smoke and particulate.

4.2.5 Injection Timing

As described earlier, the fuel is injected into the engine when the piston is close to the top of its stroke. In practice the exact timing of this injection is vital to the engine performance and emissions output.

If fuel is injected earlier, there is a tendency for more pre-mixing of fuel before combustion and for higher combustion temperatures and pressures to be developed. These conditions are in general thermodynamically advantageous and return good fuel economy and usually reduced levels of hydrocarbons, black smoke and particulate matter. However NOx emissions will be high.

If fuel is injected later there is less time available for fuel pre-mixing, pressure and temperatures in the cylinder are lower and the end of combustion occurs late in the cycle. These conditions will give low levels of NOx emission but at the expense of increased hydrocarbons, smoke, particulate matter and fuel consumption and increased exhaust temperature. The optimum timing for any given load and speed will depend on many variables:

— target emission levels;

— target fuel economy;

— engine application or duty cycle;

— maximum allowable cylinder pressure;

— maximum allowable exhaust temperature;

— fuel ignition quality;

— fuel pump injection characteristics;

— use of exhaust aftertreatment device.

4.3 *Fuel Quality*

It is important to realise that significant reductions in exhaust emissions are not attainable by fuel quality improvement alone. Many of the fundamental engine technology changes discussed in this paper, however, require that the fuel quality is maintained either at the level currently available or at an improved level.

Four features of diesel fuel quality can be considered as important with regard to exhaust emissions:

— ignition quality;

— sulphur content;

— aromatic content;

— low temperature characteristics.

4.3.1 *Ignition Quality*

This is generally measured as 'cetane number'. The higher the value of cetane number the more easily the fuel can auto-ignite when inducted into the high temperature and high pressure air in the combustion chamber. Fuel with a high cetane number will thus have a shorter ignition delay. Considering the previous discussions on the pre-mixing of fuel during the ignition delay period it will be understood that allowing ignition quality to degrade will lead to an increase in NOx emissions on a given engine. The ignition quality of a diesel fuel is partly dependent on the characteristics of the base crude oil and partly a result of refining procedures where increasing the diesel fuel yield by the addition of cracked products will reduce cetane number.

In the USA a high demand for gasoline dictates that cracked volatile components are utilised to extend gasoline yield, the less volatile components then being added to the diesel stream. This has resulted in cetane numbers having decreased from 50 to 45 in recent years.

In Europe and the rest of the world cetane numbers have been maintained at 50 or higher.

4.3.2 *Sulphur*

Sulphur is a natural constituent of crude oil, its quantity depending on the source of the oil. If not controlled during the refining process sulphur levels in diesel fuel can exceed 1% although 0.5% is typical. Sulphur in the fuel appears in the exhaust as sulphur dioxide, sulphuric acid and metal sulphates and contributes directly to particulate emissions (up to 10%).

By additional processing within the refinery (with significant implications for investment and energy consumption), sulphur levels can be reduced to very low values if required.

Specification maxima in Europe have been reduced from 0.5% to 0.3% sulphur content, with several countries requiring 0.2% in the future. In the USA, where 1994 diesel particulate legislated levels are very low, it is expected that 0.05% sulphur fuel will be available by that date. A similar value is to be required by the European Community from 1996.

4.3.3 *Aromatics*

It is believed by some that aromatic compounds (which can be up to 30% of typical diesel fuels) contribute to increased exhaust particulate levels, although the argument is complicated by interactions with other characteristics, eg ignition quality. The proportions of these aromatic compounds can be reduced by reformulation at the refinery. Examples of diesel fuel with approximately 10% aromatic content are on sale in California and Sweden.

4.3.4 *Low Temperature Operation*

Low temperature clouding and waxing of diesel fuel, together with low cetane number, will affect the ability of the engine to start and achieve a fast, clean warm-up. Under conditions of poor starting excessive levels of unburnt hydrocarbons are exhausted, primarily as visible white smoke. The waxing tendency of diesel fuel is controlled by the use of additives.

Diesel fuel sold in the UK is now acceptable for operation down to -15°C. In certain other extreme temperature markets fuel is available for operation to -40°C.

4.3.5 *Fuel Additives*

Commonly used fuel additives comprise:

— antioxidants to improve oxidation stability in storage;

— metal deactivators to provide a passive film on active metals to inhibit fuel oxidation;

— polymeric dispersants to maintain a clean fuel system;

— corrosion inhibitors to protect fuel pumps and injectors by inhibiting abrasive corrosion particles;

— biocides to inhibit biological activity and contamination;

— cetane improvers to improve ignition quality;

— low temperature flow improvers to modify the shape and size of wax crystals and to inhibit wax settling in storage tanks;

— combustion modifiers, mainly inorganic salts, to reduce smoke production and deposits in combustion chamber.

The first five categories listed above are not generally expected to have any beneficial or detrimental effects on exhaust emissions relative to fresh fuel in a clean engine. However they are generally classed as beneficial in restricting degradation of the total system. The effects modified by the last three listed (high cetane number, good low temperature operation and optimum combustion) have already been discussed. Additives in these categories would generally be expected to have beneficial effects (of magnitude dependent on the circumstances) on exhaust emissions.

Two points relating to cetane improvers should be noted. Although the cetane number, as currently measured, can be improved by additives, the performance of the fuel may not be equal to a similar fuel of equivalent but natural cetane number. It may therefore not always be satisfactory for the refinery to blend a low ignition quality fuel and raise its cetane number by the use of additives. It should also be noted that the commonly used cetane improvers are nitrate compounds which, although capable of reducing emission levels, are thought by many researchers to contribute to the carcinogenic content of the exhaust particulate.

4.4 *Lubricating Oil Additives*

The lubricating oil used in engines today is not a simple product but consists of the basestock oil and a complex package of additives. Oil additives can be considered under five broad categories:

- detergents to keep pistons clean of varnish, neutralise acidity and also to confer anti-rust properties;

- dispersants to keep insoluble material in suspension in the lubricant and to prevent flocculation and blocked oil-ways;

- viscosity improvers which give oils multi-grade viscosity characteristics, with adequate viscosity at high temperature and low viscosity at low temperature;

- anti-wear additives to reduce the severity of the effects of metal to metal contact where this occurs; they include phosphates and sulphur;

- antioxidants to inhibit oxidised products which cause oil thickening, sludge and bearing corrosion.

The additives in heavy duty diesel lubricating oil do not in general have any beneficial or detrimental effect on the level of exhaust pollutants emitted. However additive packages with reduced direct contribution to the particulate are now under active development, to assist in meeting the very severe standards of the future. Also by reducing the level of progressive degradation of the oil and the engine, a well formulated additive package will assist in maintaining the design level of emissions between oil changes and over the engine life.

It should be noted that the phosphorus content of the anti-wear additives has the potential to poison any exhaust catalyst. The additive manufacturers are aware of this and are making large reductions in phosphorus levels, without loss in anti-wear properties, by careful formulation.

4.5 *Engine Design*

Several diesel engine design features can contribute directly to the control of exhaust emissions. The most significant of these are:

- layout of combustion system;

- reduction of parasitic losses;

- features reducing oil consumption;

— enhanced air mass consumption (to give more flexibility in combustion system matching).

4.5.1 *Layout of Combustion System*

The use of two valves per cylinder (one intake, one exhaust) has always been a popular arrangement for the majority of heavy duty diesel engines. It is a layout which is relatively easy to engineer, minimises the number of components (reducing complexity and cost) and will generally enable performance targets to be achieved.

However, valve sizes are usually maximised, excluding the fuel injector from being positioned at the centre of the cylinder bore. This and other design constraints usually dictate that the fuel injector will be offset from the centre by up to 10% of the bore, and at an angle of 20-30° from the vertical.

Since the fuel is injected through several holes (typically 4 to 8 depending on the level of in-cylinder air swirl), in a radial pattern it will be understood that it is impossible, with a 2-valve arrangement, to arrange perfect symmetry of the holes in the nozzle tip with all fuel spray paths of equal length meeting identical air motion. This lack of symmetry is not of major consequence if the exhaust emissions are not controlled, but becomes significant when demanding low emissions targets are set.

In order to achieve the lowest possible exhaust emissions, compromises in the position of the fuel injector are not acceptable. For this reason the injector is best placed central to the cylinder and vertical with all fuel spray paths then truly symmetric. This positioning requires the adoption of four valves per cylinder, which is increasingly common in the most advanced engines. Other factors relating to the layout and design of the combustion chamber are also important:

— optimum size and design of the combustion bowl in the piston;

— minimising of any volume not in combustion bowl when the piston is at the top of its stroke (minimum piston to head clearance, avoidance of valve cutouts in piston and minimised valve recession, high position top piston ring);

— optimum match of swirling air motion and available fuel pressure (as fuel pressures increase, air swirl can be advantageously reduced).

4.5.2 *Reduction of Parasitic Losses*

Parasitic losses can be considered to be:

— mechanical friction;

— gas pumping losses (filling, compression and emptying of gases);

— liquid pumping losses (coolant, lubricant).

Reduction in parasitic losses has two benefits:

— all lost or wasted work from the fuel will increase fuel consumption and directly add to the level of undesired exhaust emissions;

— reducing the losses will enable the engine to redirect the original 'lost' power as useful work resulting in either more power for the same emissions output, or less emissions for the same power output.

Legislation limits for exhaust emissions are set in 'specific power' units (grams/kW hour). An increase in power as the result of reducing parasitic losses, whilst emissions remain constant, will therefore reduce the certified emission level.

The minimising of parasitic losses is most effectively undertaken during the design of a new engine when all best practices can be incorporated, by careful attention to detailed engineering.

Major sources to be addressed include:

    — piston friction – ring design, piston profile;

    — oil and water pump efficiency;

    — breathing losses and optimisation of compression ratio;

    — optimisation of turbocharger specification.

5. Emissions After-Treatment Control Systems

After all practical steps have been taken to inhibit, at source, the formation of undesired pollutants, further reductions will require that the emissions be chemically or mechanically removed from the exhaust stream (after-treatment).

5.1 Catalytic Converter

Petrol powered passenger cars have for several years in the USA, and more recently in parts of Europe and Scandinavia, been equipped with catalytic converters in the exhaust system. Within the converters the exhaust gas passes through an expanded medium (typically ceramic or metal) which presents to the exhaust a large surface area coated with a suitable catalyst (usually platinum, palladium, rhodium and other trace elements). The catalysts promote oxidation of CO and hydrocarbons to CO_2 and H_2O (2-way oxidising catalysts) and may additionally promote reduction of NOx to N_2 and O_2 (3-way catalysts) once a 'light off' temperature is reached in the exhaust system. The presence of excess oxygen in the diesel exhaust stream is fundamental and prohibits the use of a reducing catalyst for NOx control, with existing catalyst technology. Since diesel CO emissions are relatively low the use of a catalytic converter is primarily for the oxidation of hydrocarbons to control gaseous hydrocarbons and help reduce particulate emission levels. However diesel exhaust temperatures are very low at light load, due to the large excess of air, and the catalyst will not then light off.

An unwanted associated reaction is the tendency for sulphur dioxide in the exhaust to oxidise to sulphur trioxide which forms sulphuric acid with water. Catalysts are being designed to limit this conversion, but as fuel sulphur levels are progressively decreased the problem will become less significant.

The catalytic filter will decrease exhaust particulate levels under operating conditions where a high proportion of the particulate is liquid or gaseous hydrocarbons. This suits it well for mixed load operation such as experienced by the city bus in preference to long periods of high load operation such as with very large vehicles.

Items in favour of catalyst use are:

    — effective control of liquid and gaseous hydrocarbons, CO;

    — effective control of volatile components of particulates;

— no control system is required;

— proven technology;

— reasonable cost.

Items against catalyst use are:

— a limited life expectancy – the catalyst can age, or be 'poisoned' by undesirable products in the fuel;

— a liability to unauthorised removal or engine operation when the catalyst has failed.

5.2 *Particulate Trap*

Where levels of exhaust particulate are required below that attainable by engine improvements and catalyst control it becomes necessary to consider other means of after-treatment.

Since the residence time of particulate matter in a free flowing exhaust system is insufficient to permit significant oxidation to take place, even in the presence of catalysts, it is necessary to trap the particles before they are exhausted into the atmosphere and to store them until they can be disposed of. Various methods of collecting exhaust particulate have been evaluated including centrifuges, water scrubbers, electrostatic precipitators, oil baths and filters. Of these only the filter has been shown to be practicable, with other methods failing to achieve sufficient trapping efficiency.

Any 'simple' filter would rapidly become blocked as it continued to trap the solid particulate matter and would eventually present an unacceptably high exhaust back pressure to the engine. This would seriously affect engine performance and result in increased levels of emissions and fuel consumption. It is therefore necessary to clear the trap of particulates before blocking becomes a problem.

Consideration of the amount of particulate matter required to be trapped reveals that the storage volume required for a throwaway or service cleanable trap is impractical for automotive applications. Many studies of the problem have concluded that the only feasible solution is to arrange for the particulate matter to be oxidised or combusted to clean the filters *in situ*. This process is known as trap regeneration.

After several years of development most of the problems preventing application of particulate traps have been overcome. A large variety of particulate trap systems are under advanced development in Europe and the USA both by independent trap manufacturers and by the engine producers, with trapping efficiencies of 70 – 80% commonly observed. Some prototypes are currently undergoing fleet trials in order to evaluate their suitability for introduction into limited sections of the market.

The following features are usually present:

— a monolithic ceramic filter;

— energy for regeneration supplied either electrically or, more often, from a diesel fuel operated burner;

— sophisticated control systems to monitor build up of exhaust back pressure, initiate and maintain regeneration and control exhaust flow as required.

Although there is currently no strong technical argument against the use of particulate traps, the financial and operational considerations are of concern. With their expensive filter medium, regeneration system and sophisticated control system, current traps account for a high proportion of the total engine cost and as presently developed will require replacement several times during the life of the engine.

5.3 Exhaust Gas Recirculation

This is a technique whereby a proportion of the exhaust gas is returned to the combustion chamber displacing some of the fresh air charge from the next cycle. This reduces the peak combustion temperature and hence reduces the level of NOx produced. Exhaust gas recirculation (EGR) is usually achieved by means of a suitably controlled valve. EGR has been shown to be an effective means of reducing NOx, usually showing less of a rise in hydrocarbons than is seen by achieving the same NOx levels with retarded injection time.

There are, however, serious side effects:

— the recycling of soot can affect the performance of the intake system;

— carbon particles are hard and recycling can increase cylinder wear.

— the exhaust gas displaces air, reducing oxygen availability and tending to give worse levels of CO, smoke and particulate at high load conditions. This dictates that EGR levels must be carefully controlled.

5.4 NOx Control Aftertreatment

Several researchers have developed processes for the purpose of decreasing NOx levels in the exhaust system. Two processes for which limited results from test demonstrations have been published are:-

Raprenox (Rapid Removal of NOx) – This system is based on the use of crystallised cyanuric acid, a harmless compound used in swimming pools to stabilise chlorine levels. The temperature of the acid crystals is raised above 330°C in an exhaust heated chamber, at which temperature isocyanic acid gas (HNCO) is released. The gas is mixed with the engine exhaust where, at temperatures above 400°C, a series of reactions occur which result in the removal of oxides of nitrogen, leaving carbon monoxide, carbon dioxide and water. Claims have been made for NOx removal rates in excess of 90%.

SCR (Selective Catalytic Reduction) – The SCR process uses a ceramic molecular sieve catalyst. Ammonia (NH_3) is injected into the engine exhaust system and together with the oxides of nitrogen enter the micro-pore structure of the molecular sieve. The exothermic reaction of NOx with NH_3 in the small pore space generates nitrogen gas and water which are expelled from the pore structure back into the gas stream. The quantity of ammonia injected is dependent on the level of NOx to be removed. The catalyst temperature must be maintained between 300-400 C. NOx reduction in excess of 90% has been claimed.

Both the above processes require at least:

— an on board supply of reducing chemical;

— a reaction chamber;

- microprocessor control of metered chemical supply and reaction temperature;

- monitoring of NOx levels.

Health effect implications of both systems are currently unknown.

There is no doubt that NOx can be successfully removed from the exhaust stream but the costs and complexity of the systems demonstrated so far are high. Although such systems have been used for stationary engines, interest for mobile sources has been minimal. This reflects the cost sensitivity of this sector which at present would not be willing to accept the cost and packaging implications of both NOx removal and particulate trapping technology.

6. Electronics

In order to meet future levels of exhaust emissions and to achieve required standards of fuel economy, ease of driving, cold starting etc, the extensive use of electronics will be implicit. The economics of the overall design will dictate that the number of electronic packages will be minimised, the ultimate solution being where one microprocessor would control all engine, vehicle and associated equipment functions.

6.1 *Engine Management*

The widespread adoption of electronics will enable several parameters to be controlled accurately and with a flexibility not possible with mechanical control. Features which will benefit are:

- injection timing – enabling close control of the timing at all loads and speeds, to best meet the requirements of the test cycle and the engine's in-service duty cycle;

- injection rate – control of the mechanical/hydraulic mechanism of varying injection rate to optimise combustion requirements;

- variable geometry turbocharging – enabling accurate control of the turbocharger for optimum boost levels at all engine speeds and loads, and during transient operation;

- exhaust gas recirculation – adjustment of levels to suit the requirements of each operating point;

- variable valve timing – as complex mechanical solutions are required the best control systems will use electronics. The potential benefits, however, are small on heavy duty diesel engines.

- governing – control of a heavy duty diesel engine is usually by means of a mechanical governor. This task could be performed by electronic control although safety requirements will have to be met in the event of system malfunction.

- transient control – the current European 13-mode test requires optimising emission output at steady loads and speeds. In order to minimise environmental impact it becomes necessary to maintain optimum control of all the above variables through rapid engine load and engine speed changes.

6.2 *Diagnostics*

Diagnostic tools enable the performance and deterioration of any chosen variable to be monitored. Given the complexity and precision of engineering of future low emissions engines, considerably more sophisticated means of checking the performance of the engine will be required than is provided by the subjective approach common today in vehicle operation. Additionally, emissions levels will be so low in the future that deterioration will not be immediately obvious even to trained observers (eg visible smoke will be non-existent, even when particulates are outside legislative limits). In-service checks therefore will be difficult to implement cost-effectively and may not be feasible on some pollutants.

Considering the wide array of complex and expensive equipment in engine test laboratories for the measuring and certification of engine emission levels, it will be appreciated that such a system could not be applied as an on-board vehicle diagnostic system.

However other features of the engine operation can be monitored and judged against design requirements to indicate overall health. Many of these can be integrated with the electronic management system:

— engine speed;

— fuel injection timing and duration;

— turbocharger speed;

— inlet and exhaust temperature and pressure;

— maximum cylinder pressure;

— blowby gases escaping past the piston rings;

— coolant and oil temperature and pressure.

Much effort has been expended in recent years researching 'intelligent' diagnostic systems. These monitor factors which can be directly measurable, and use any noted changes to predict the behaviour of more complex factors. The performance of such systems varies from fairly simple to highly intelligent. It is likely that intelligent systems will grow progressively out of electronic diagnostics and not at any stage be 'introduced'.

7. Retrofitting
Retrofitting is generally considered to include the rebuilding of old engines to 'modern' standards, and the fitting of bolt-on devices. The gains in reduction of emissions must be weighed against practicality and cost.

7.1 Rebuilding
Unlike the light duty passenger car engine, in some applications the heavy duty diesel engine may be rebuilt (usually referred to as a major overhaul) during its lifespan, since particular components (eg cylinder liners, piston rings, bearings) have a shorter life to wear-out than structural components (eg cylinder blocks, cylinder heads). Typical applications where this occurs are urban buses and some long-haul trucks. During the rebuild there is some scope to 'modernise' the engine to reduce emissions by fitting upgraded components but the cost of the rebuild may increase significantly. However at very low emissions levels (as targeted for the middle of the present decade) the engineering of the engine becomes a complex integrated package, such that a rebuild is approaching replacement by a new engine. This therefore raises further the cost-effectiveness of maintenance workshop rebuilding, as against an engine-swap programme set up by the manufacturer. The latter would always be technically feasible, and would ensure the replacement engines met certification standards, but would only become widespread if made financially attractive to the vehicle owner/operator.

7.2 *Bolt-on Devices*

Means of improving the emissions output from existing engines are primarily:

— particulate traps – an effective but expensive solution to improve particulate levels from existing engines. It should be noted that the 'old' engine may be a high base emitter of particulates, and could require a higher capacity filter system than the equivalent 'new' engine;

— catalytic converters – these would offer a relatively low cost, easily supplied and effective reduction in levels of hydrocarbons and carbon monoxide;

— diesel fuel improvements – not essentially a retrofit, but it should be remembered that improvements in diesel quality (ie the growing presence of 'green' fuels) with low aromatic content, high ignition quality, low sulphur and good cold operation will improve the particulate emission output from 'old' engines.

8. Consequences of Design Changes

The main demands which have influenced heavy duty diesel design in recent years and which remain of influence are:

— increasingly lower exhaust emissions;

— improvements in fuel efficiency;

— reductions in noise and vibration;

— increase in power density with more power from smaller engines;

— increase in service intervals;

— maximised reliability and durability;

— minimised total life costs.

8.1 *Impact of Low Emission Technologies*

The technologies discussed in this paper are being ever more widely adopted for heavy duty diesel engines as the low emissions requirements continue to be increasingly demanding. These technologies will also have side effects:

— higher fuel injection pressures – increases cost of fuel injection equipment;

— retarded injection timings – increases exhaust temperatures reduces fuel efficiency, reduces combustion noise;

— Four valves per cylinder, central injector – increases complexity and therefore cost, increases mechanical noise sources;

— lower air charge temperatures through more efficient aftercoolers – reduces thermal loading on engine components;

— reductions in maximum engine speed – reduces frictional losses but require increased mechanical loading for equal power output;

— improved fuel quality – will generally benefit combustion noise levels and fuel efficiency;

— improved lubricant quality – will generally increase change intervals, reduces wear rates;

— exhaust aftertreatment – increase in cost through reduced fuel economy and increased service requirements;

— electronics – increased cost in the short term but potentially lower costs in the long term with potential benefits to fuel economy, noise control and service requirements.

9. Meeting Emission Limit Value Requirements

9.1 *Impact of Production Realisation*

Engineering constraints dictate that it is necessary to design an engine so as to achieve better than a given legislated emission level. Factors producing and controlling NOx levels are in general fairly controllable, such that to aim to be 10% better than the desired limit may be considered acceptable. Factors which produce and control particulate output are in general more variable, such that to engineer the engine to be at least 20% better than the desired limit would usually be required.

9.2 *Timing of Production Realisation*

Except where stated otherwise the technologies discussed in this paper are basically available for implementation at present. They have been developed with the objective of being compatible with manufacturing requirements of the engine builder and the durability and reliability demands of the end user.

However, once a particular technology is identified as a requirement for a given engine to meet a given emission limit, a significant period will elapse before the modified engine can be released to the market. That period results from the need to prepare and modify manufacturing techniques and systems, certify the new specification and prepare the appropriate dealer and after-service procedures.

It is to be expected that relatively simple modifications to an existing engine (such as revised fuel injection equipment, modified combustion system etc) take at least two years for introduction. Where major modifications to the base engine are required it is expected that this period of introduction is likely to be in excess of four years.

GLOSSARY

Terms printed in *italics* have a separate entry in the glossary.

Aftercooling (of air)	The cooling of air compressed in a *turbocharger*, before it enters the engine cylinders, so as to increase its density.
Aftertreatment (of exhaust)	Application of a component system to the exhaust of an engine, to control the pollutants in the exhaust gas stream.
Aromatics	Compounds whose molecules contain carbon atoms (and in some cases other atoms) arranged in one or more benzene rings with double bonds between some of the atoms. They are more dense than most of the other components of diesel fuel and have higher boiling points.
Carcinogen (chemical)	A substance that is capable of increasing the tumour burden of a mammal. Carcinogenesis, in contrast to mutagenesis, is a comparatively rare change which occurs after a latency period whose length is related to the life span of the organism. It may sometimes become evident only after the affected individual has reproduced. Some chemicals which are not *mutagens* can contribute to the development of tumours, by mechanisms that are not yet fully understood.
Catalytic trap	A *particulate trap* in which the filter element is coated with a catalyst to promote combustion of the trapped *particulate* matter at a normal exhaust temperature.
Certification	The process of testing, in this case an engine, to demonstrate compliance with set standards of performance. It is carried out at approved premises under specified and closely controlled conditions, often witnessed by a representative of the certifying authority.
Cetane index	A combined measure of the boiling point(s) and density of diesel fuel, used to monitor its production in the refinery. It gives a good approximation to the *cetane number* of the fuel before the addition of any ignition improver.
Cetane number	The measure of the ease of ignition of a diesel fuel, that is, the delay between injection and ignition in an engine. It is measured by a standard procedure on a specified test engine, by comparison with fuel mixes of known ignition quality.
Charge (of air)	The mass of air taken into an engine's cylinders on each cycle.
Conformity of production	An aspect of some *certification* procedures, designed to ensure that engines in volume

production meet the standard set. The standard may be somewhat less tight than that required for *type approval.*

Dynamometer	A machine which is coupled to an engine on test, so as to absorb the shaft power developed and to allow the *torque* produced by the engine to be measured.
Exhaust gas recirculation	A method of control of NOx emissions in which a proportion of the exhaust gas is mixed with the incoming air so as to reduce the average temperature of gas in the cylinder during combustion.
Flow-through catalyst	An exhaust *aftertreatment* system employing a catalytic membrane which promotes chemical reactions in the exhaust gas stream without trapping particulate material. It is used mainly to oxidise unburnt hydrocarbons into carbon dioxide and water.
Fuel injection	The process by which fuel is introduced into an engine's cylinders, in the form of one or more finely atomised sprays. In a diesel engine the design and operation of the fuel injection system are major factors in the formation and combustion of the fuel/air mixture.
Hydrocarbons (HC)	Compounds of hydrogen and carbon. The main constituent of diesel fuel and of lubricating oil. Hydrocarbons in the exhaust gas stream result from incomplete combustion of fuel or lubricating oil. They may be gaseous or in the form of condensed droplets of liquid or be adsorbed onto solid material.
Limit value	The value specified in regulations for the maximum permissible emissions of a given pollutant. Different values may be set for *type approval* and *conformity of production.*
Load	For an engine, the resistance which is overcome by the *torque* delivered. Numerically equivalent to torque.
Mutagen (chemical)	A substance that is capable, either of itself or through a metabolite, of modifiying the genome of an organism by altering its DNA. Such an alteration may contribute to the development of cancer after a latent period and may have other adverse effects.
Naturally aspirated (engine)	An engine without any kind of air compressor feeding it, so that (in a 4-stroke configuration) the piston motion on the idle stroke is alone responsible for drawing a fresh *charge* of air (or air/fuel mixture) into the cylinder. See also *turbocharging.*
Nitrogen oxides (NOx)	A generic term for the oxides of nitrogen and for a mixture of them. In an engine, they are formed by

direct reaction between nitrogen and oxygen in the air, heated by combustion of the fuel, and by further oxidation in the exhaust stream.

Opacity	A measure of the visual 'density' of exhaust emissions. Can be tested by passing a beam of light through the exhaust plume and observing the extent to which it is attenuated, per metre of width of the plume.
Particulate emissions or particulates	Fragments of carbon or other solid matter emitted from an engine, often with adsorbed gaseous components, and droplets of liquid. For the purposes of engine *certification* it is defined as all material which, after cooling and dilution of the exhaust gas stream, is trapped on a filter under specified conditions of sampling.
Particulate trap	An exhaust *aftertreatment* system, often incorporating a porous ceramic block, designed to filter and store the *particulates* in the exhaust gas stream. To prevent blockage it requires *regeneration*.
Polyaromatic hydrocarbons (PAHs)	A class of *aromatics* in which the molecules contain several linked benzene rings of atoms. Some PAHs are *carcinogens*.
Power	The rate of doing work. For an engine, the output measured in kilowatts or brake horse-power. It is calculated as the product of *torque* and engine rotational speed.
Regeneration	The process applied to the filter element of a *particulate trap* to clean it of stored *particulates*, generally by combustion. The combustion may be triggered by electrical or flame heating or by *catalytic* means.
Retrofit	The process of modifying an engine or vehicle to a new specification or of adding a system such as exhaust *aftertreatment*.
Smoke, black	Visible smoke consisting of *particulates*. In an engine, it arises from pyrolysis and incomplete combustion of fuel, generally under conditions of high *load*.
Smoke, white	From an engine, an aerosol of partially or totally unburnt fuel. It is normally emitted only when the engine itself, and the ambient air, are cold. Lubricating oil may also be a source of white smoke in a worn engine.
Torque	Turning effect or moment; the equivalent, in rotation terms, to linear force. The output from an engine. See also *load* and *power*.
Turbocharging	The most commonly used of several possible techniques for compressing air before it enters the engine cylinders, so as to increase the mass of air available for combustion and hence to increase the

power output of the engine. Such compression increases the temperature of the air. See also *aftercooling*.

Type approval

The *certification* of an engine type or vehicle specification to a set standard by testing a representative engine or vehicle. See also *conformity of production*.

REFERENCES

1. ROYAL COMMISSION ON ENVIRONMENTAL POLLUTION (1984). *Tackling Pollution — Experience and Prospects:* Tenth Report. **Cmnd. 9149.** HMSO, London.

2. DEPARTMENT OF TRANSPORT, SCOTTISH DEVELOPMENT DEPARTMENT, WELSH OFFICE (1990). *Transport Statistics Great Britain 1979-1989.* HMSO, London.

3. DEPARTMENT OF TRANSPORT. Evidence to the Royal Commission.

4. EC DIRECTIVE (1987). Council Directive on the Approximation of Laws of the Member States Relating to Measures to be Taken against the Emission of Gaseous Pollutants from Diesel Engines for Use in Vehicles, 88/77/EEC. *Official Journal of the European Communities* **L36,** 33.

5. COUNCIL OF THE EUROPEAN COMMUNITIES (1991). Common Position Adopted by the Council on 13 May 1991 with a View to the Adoption of the Directive Amending Directive 88/77/EEC. *Council Paper 5701/91.*

6. EC COUNCIL DIRECTIVE (1984). Council Directive 84/424/EEC of 3 September 1984 Amending Directive 70/157/EEC on the Approximation of the Laws of the Member States Relating to the Permissable Sound Level and Exhaust System of Motor Vehicles. *Official Journal of the European Communities* **L238,** 31.

7. UK QUIET HEAVY VEHICLE PROJECT; a collaborative project between UK Government and industry. Report to be published.

8. PERKINS TECHNOLOGY LTD. Evidence to the Royal Commission.

9. H M GOVERNMENT (1990). *This Common Inheritance — Britain's Environmental Strategy.* **Cmnd. 1200.** HMSO, London.

10. SECRETARY OF STATE FOR TRANSPORT (1991). Speech to a conference, 28 May 1991.

11. NEEDHAM J.R., DOYLE D.M., FAULKNER S.A. (1990). *Developing the Truck Engine for Ultra Low Emissions.* Presented at a seminar of the Institute of Mechanical Engineers, 1 February 1990.

12. FELLOWSHIP OF ENGINEERING. Evidence to the Royal Commission.

13. LUBRIZOL CORPORATION. Evidence to the Royal Commission.

14. FREIGHT TRANSPORT ASSOCIATION. Evidence to the Royal Commission.

15. HOPE K. (1991). Greek Bus Stops Smog in its Tracks. *Financial Times,* 25 June 1991.

16. UMWELTBUNDESAMT (Federal Environmental Agency, Germany). Evidence to the Royal Commission.

17. ENGINE MANUFACTURERS ASSOCIATION, USA. Evidence to the Royal Commission.

18. DEGUSSA AG. Evidence to the Royal Commission.

19. DAF BV. Evidence to the Royal Commission.

20. TRANSPORT AND ROAD RESEARCH LABORATORY (1990). UK Road Transport's Contribution to Greenhouse Gases; a review of TRRL and other research. *TRRL Contractor Report* **223.** HMSO, London.

21. WORLD WIDE FUND FOR NATURE. Evidence to the Royal Commission, building on work published as *Atmospheric Emissions from the Use of Transport in the United Kingdom* (1989). A Report for the World Wide Fund for Nature prepared by Earth Resources Research Ltd.

22. DEPARTMENT OF ENERGY (1989). *Digest of United Kingdom Energy Statistics 1989.* HMSO, London.

23. DEPARTMENT OF THE ENVIRONMENT (1990). *Digest of Environmental Protection and Water Statistics.* HMSO, London.

24. WARREN SPRING LABORATORY (1990). *National Atmospheric Emissions Inventory.* Warren Spring Laboratory, Stevenage.

25. UNITED KINGDOM PHOTOCHEMICAL OXIDANTS REVIEW GROUP (1990). *Oxides of Nitrogen in the United Kingdom:* Second Report. Department of the Environment, London.

26. ATOMIC ENERGY RESEARCH ESTABLISHMENT (1980). Ozone Precursor Relationships in the United Kingdom. *AERE Report* **R12408.** HMSO, London.

27. ADVISORY GROUP ON THE MEDICAL ASPECTS OF AIR POLLUTION EPISODES (1991). *Ozone:* First Report. Department of Health, London.

28. HOUGH A.M., DERWENT R.G. (1990) Changes in the Global Concentration of Tropospheric Ozone due to Human Activities. *Nature* **344,** 645-648.

29. MANSFIELD T.A. *The Environmentalist* (In press).

30. DEPARTMENT OF THE ENVIRONMENT. Evidence to the Royal Commission.

31. MANSFIELD T.A. (1989) *The Soiling of Buildings in Urban Areas.* Ph.D. Thesis. Middlesex Polytechnic, Centre for Urban Pollution Research.

32. EC DIRECTIVE (1985). Council Directive on Air Quality Standards for Nitrogen Dioxide, 85/203/EEC. *Official Journal of the European Communities* **L87,** 1.

33. EC DIRECTIVE (1980) Council Directive on Air Quality Limit Values and Guide Values for Sulphur Dioxide and Suspended Particulates, 80/779/EEC. *Official Journal of the European Communities* **L229,** 30.

34. STATUTORY INSTRUMENT (1989). Clean Air, The Air Quality Standards Regulations 1989. *S.I. 1989,* **317.** (Similar Regulations came into force for Northern Ireland in 1990.)

35. WORLD HEALTH ORGANISATION (1987). Air Quality Guidelines for Europe. *WHO Regional Publications, European Series,* **23**.

36. INTERNATIONAL AGENCY FOR RESEARCH ON CANCER (1989). Diesel and Gasoline Exhausts and Some Nitroarenes. *IARC Monographs on the Evaluation of Risks to Humans* **46**. World Health Organisation, Geneva.

37. GUSTAVSSON P., PLATO N., LIDSTROM E., HOGSTEDT C. (1990) Lung Cancer and Exposure to Diesel Exhaust among Bus Garage Workers. *Scandinavian Journal of Work and Environmental Health* **16**, 348-354

38. SWEDISH GOVERNMENT COMMITTEE ON AUTOMOTIVE AIR POLLUTION (1983). *Motor Vehicles and Cleaner Air.* Stockholm.

39. DEPARTMENT OF HEALTH. Evidence to the Royal Commission.

40. WORLD HEALTH ORGANISATION (1990). *Impact on Human Health of Air Pollution in Europe.* WHO Regional Office for Europe, Copenhagen.

41. WEBB J. (1991). Car Exhausts May Cause Hay Fever. *New Scientist* **130**, (1774) 22.

42. READ R.C., GREEN M. (1990). Internal Combustion and Health. *British Medical Journal* **300**, 761-762.

43. VAN DEN HOUT K.D., RIJKEBOER R.C. (1986) Diesel Exhaust and Air Pollution. *Research Institute for Road Vehicles Report* **R86/038.** TNO, Delft.

44. BALL D., CASWELL R. (1983). Smoke from Diesel Engine Road Vehicles: an Investigation into the Basis of British and European Emission Standards. *Atmospheric Environment* **17**, 169-181.

45. NEWBY P.T., MANSFIELD T.A., HAMILTON R.S. (1991). The Economic Implications of Building Soiling in Urban Areas. *Science of the Total Environment* **100**, 347-365.

46. HECK W.A., ADAMS R., CURE W.W., HEAGLE A.S., HEGGESTAD H.E., KOHUT R.J., KRESS L.W., RAWLINGS J.O., TAYLOR O.C. (1983). A Reassessment of Crop Loss from Ozone. *Environmental Science and Technology* **17**, 573A-581A.

47. INTERGOVERNMENTAL PANEL ON CLIMATE CHANGE (1990). *Climate Change*. Cambridge University Press, Cambridge.

48. UNITED KINGDOM REVIEW GROUP ON ACID RAIN (1990). Acid *Deposition in the United Kingdom 1986-1988:* Third Report. Department of the Environment, London.

49. JONES K.C., STRATFORD J.A., TIDRIDGE P., WATERHOUSE K.S. (1989). Polynuclear Aromatic Hydrocarbons in an Agricultural Soil: long-term changes in profile distribution. *Environmental Pollution* **56**, 337-351.

50. STATUTORY INSTRUMENT (1990). The Road Vehicles (Construction and Use) Regulations 1990. *S.I. 1990*, **1131**.

51. STATUTORY INSTRUMENT (1986). The Road Vehicles (Construction and Use) (Amendment No.2) Regulations 1986. *S.I. 1986*, **1078**.

52. EC DIRECTIVE (1972). Council Directive on the Approximation of the Laws of the Member States Relating to Measures to be Taken against the Emissions of Pollutants from Diesel Engines for Use in Vehicles, 72/306/EEC. *Official Journal of the European Communities* **L 190**, 1.

53. BRITISH STANDARDS INSTITUTE (1971). Specification for the Performance of Diesel Engines for Road Vehicles. **BS AU 141a**.

54. COMMISSION OF THE EUROPEAN COMMUNITIES (1990). Proposal for a Council Directive amending Directive 88/77/EEC. COM(90) 174 final.

55. FRIENDS OF THE EARTH. Evidence to the Royal Commission.

56. WALSH M.P. (1989). Worldwide Developments in Motor Vehicle Diesel Particulate Control. *Proceedings of the 8th World Clean Air Congress, 1989*, **4**, 437-440. Elsevier, Amsterdam. Quoted by the National Society for Clean Air and Environmental Protection in evidence to the Royal Commission.

57. COMMITTEE OF COMMON MARKET AUTOMOBILE CONSTRUC-TORS (CCMC). Evidence to the Royal Commission.

58. CENTRAL COUNCIL FOR ENVIRONMENTAL POLLUTION CON-TROL, JAPAN (1989). *On the Future Policy for Motor Vehicle Exhaust Emission Reduction*. Environment Agency, Tokyo.

59. COMMISSION OF THE EUROPEAN COMMUNITIES. Evidence to the Royal Commission.

60. LUCAS DIESEL SYSTEMS. Evidence to the Royal Commission.

61. SAAB-SCANIA AB. Evidence to the Royal Commission.

62. VOLVO TRUCK CORPORATION. Evidence to the Royal Commission.

63. NETHERLANDS GOVERNMENT (1990). *Emissions Test Procedure for Heavy Duty Diesel Engines*. A presentation to the Motor Vehicle Emissions Group of the European Commission. Submitted in evidence to the Royal Commission by the Umweltbundesamt, Germany.

64. UNITED STATES ENVIRONMENTAL PROTECTION AGENCY. Evidence to the Royal Commission.

65. J M DUNNE. Communication with the Royal Commission.

66. D BROOME. Consultant to the Royal Commission.

67. WEST GLAMORGAN COUNTY COUNCIL. Evidence to the Royal Commission.

68. EC DIRECTIVE (1977). Council Directive on the Approximation of the Laws of Member States Relating to the Measures to be Taken against the Emission of Pollutants from Diesel Engines for Use in Wheeled Agricultural or Forestry Tractors, 77/537/EEC. *Official Journal of the European Communities* **L220**, 38.

69. INTERNATIONAL STANDARDS ORGANISATION (1991) Reciprocating Internal Combustion Engines — measurement of Exhaust emissions: Part 3; Exhaust Smoke — definition of methods of measurement under steady conditions. **ISO TC 70–SC8–N6.**

70. NATIONAL SOCIETY FOR CLEAN AIR AND ENVIRONMENTAL PROTECTION. Evidence to the Royal Commission.

71. DR M F FOX. Evidence to the Royal Commission.

72. DEPARTMENT OF FUEL AND ENERGY, LEEDS UNIVERSITY. Evidence to the Royal Commission.

73. CUMMINS ENGINE COMPANY LTD. Evidence to the Royal Commission.

74. HOLMAN C.D. (1990). FEAT (Fuel Efficiency Automobile Test): remote sensing of vehicle exhaust emissions. *Clean Air* **21**, 27-31.

75. DERBY CITY COUNCIL. Evidence to the Royal Commission.

76. VOLVO TRUCKS (GB) LTD. Evidence to the Royal Commission.

77. SHELL INTERNATIONALE PETROLEUM MIJ. Evidence to the Royal Commission.

78. MOTOR VEHICLE EMISSIONS GROUP (1990). *Fuel Quality and Diesel Emissions.* Report of a sub-group of the Motor Vehicle Emissions Group of the European Commission. Submitted in evidence to the Royal Commission by the UK Petroleum Industry Association.

79. CONCAWE, the oil companies' European organization for environmental and health protection. Evidence to the Royal Commission.

80. COMMISSION OF THE EUROPEAN COMMUNITIES (1991). Proposal for a Council Directive Relating to the Sulphur Content of Gasoil, COM(91) 154 final. *Official Journal of the European Communities* **C 174**, 18.

81. BRITISH STANDARDS INSTITUTE (1988). Specification for Automotive Diesel Fuel (Class A1). **BS 2869, Part I.**

82. DEPARTMENT OF ENERGY. Evidence to the Royal Commission.

83. UNITED KINGDOM PETROLEUM INDUSTRY ASSOCIATION. Evidence to the Royal Commission.

84. J A TERNING. Communication with the Royal Commission.

85. CALIFORNIA AIR RESOURCES BOARD (1988). Proposed Adoption of Regulations Limiting the Sulphur Content and Aromatic HydroCarbon Content of Motor Vehicle Diesel Fuel.

86. ROAD HAULAGE ASSOCIATION. Evidence to the Royal Commission.

87. TECHNICAL COMMITTEE OF THE PETROLEUM ADDITIVE MANUFACTURERS IN EUROPE (ATC). Evidence to the Royal Commission.

88. PARAMINS — EXXON CHEMICAL TECHNOLOGY CENTRE. Evidence to the Royal Commission.

89. NEEDHAM J.R., DOYLE D.M., FAULKNER S.A. FREEMAN H.D. (1989). Technology for 1994. *SAE Technical Paper* **891949**.

90. MAN NUTZFAHRZEUGE AG (1990). Prospects for Alternative Fuels? *MAN Magazine* **3**, 27–31.

91. BOROUGH OF BLACKBURN. Evidence to the Royal Commission.

92. CHLORIDE SILENT POWER LTD. Evidence to the Royal Commission.

93. UNITED KINGDOM PHOTOCHEMICAL OXIDANTS REVIEW GROUP (1987). *Ozone:* First Report. Department of the Environment, London.

INDEX

Printed in the United Kingdom for HMSO
Dd.0506934, 8/91, C20, 518240,5673, 162239